METRICS

Measurement for Tomorrow

METRICS

Measurement for Tomorrow

BY

HELMER A. RONNINGEN

COLLIER BOOKS
A Division of Macmillan Publishing Co., Inc.
NEW YORK

COLLIER MACMILLAN PUBLISHERS
LONDON

Metrics **was originally published in a hardcover edition under the title *Weights and Measures: U.S. to Metric—Metric to U.S.* by Ronningen Metric Company and is reprinted by arrangement.**

Library of Congress Catalog Card Number: 72-77653

First Collier Books Edition 1972
Fourth Printing 1976

Designed and Illustrated by Ray Prather

Macmillan Publishing Co., Inc.
866 Third Avenue, New York, N. Y. 10022
Collier Macmillan Canada, Ltd.

Printed in the United States of America

SIMPLE "RULES FOR THE ROAD"

A liter is a quart–

but add a little

A kilogram is two pounds–

but add a little

A kilometer is ½ mile–

but add a little

100 grams is ¼ pound of candy–

but add a little.

CONTENTS

PREFACE

As the owner and controller of a machine manufacturing concern, I have traveled extensively throughout North America, Europe, Asia, and the world. I have furnished machinery and equipment to paper mills, chemical plants, and general industry and several of the designs we produced received world-wide patents. In order to sell around the world it was necessary to adapt the machinery for the Metric System, and, to aid the customer, conversion tables were printed and distributed. This booklet is now available to all those who use either the U.S. or the Metric System. The conversions are made in figures and charts, with the thought that charts make for simple, easy conversion. However, if accuracy to the third decimal place is necessary, computations should be made.

The purpose of this booklet is to offer the relationship between U.S. and Metric standards of weights and measures in as simple a graph form as possible combined with arithmetical computations. This will assist laymen, students, and engineers in easily and quickly converting U.S. to Metric and vice-versa.

The U.S. system goes back thousands of years. It was developed by the Egyptians, Greeks, and Romans and was brought to England by Caesar's legions. An inch was the width of a soldier's thumb, a foot the length of his boot—

which closely figured twelve inches to the foot. A cubit was from the elbow to the fingertip, two cubits was a yard or about three feet. A mile was 1,000 paces of two steps, or 5,000 feet (this mile later became 5,280 feet or 1,760 yards).

One grain was a grain of wheat, one pound was 7,000 grains or 16 ounces. A year was considered 360 days (now 365¼), or 360 revolutions of the earth. One earth revolution was 24 hours, one hour 15° of revolution or 60 minutes or 3,600 seconds.

On June 14, 1836, Congress adopted the U.S. Standard of Weights and Measures. Each state was furnished, by the Federal government, with a complete set of weights and measures. The fundamental units were:

1 yard = 36 inches as measured between the 27th and 63rd inches of a certain 82-inch brass bar
1 pound = 7,000 grains
1 gallon = 231 cubic inches
1 bushel = 2,150.42 cubic inches

In the meanwhile, many European countries had decided upon a decimal system, known as the Metric System of Weights and Measures.

1 liter is the volume of pure water—weighing 1 Kilogram at 4° C.

1 meter is defined as a length of a platinum-iridium graduated X cross section bar.

In May, 1875, in Sèvres, France, seventeen countries including the United States agreed upon one—and one only—permanent International Bureau of Weights and Measures and established:

1 International Prototype Kilogram
1 International Prototype Meter

In April, 1893, T. C. Mendenhall, U.S. Superintendent of Weights and Measures, with approval of the Secretary of the Treasury, notified all state governments that the International Prototype Kilogram and the International Prototype Meter would be regarded as the fundamental standards of mass and length in the United States; that all states were authorized to utilize the Metric System. This was not obligatory, U.S. weights and measures would continue as such—and this is still in effect.

In 1901, the U.S. National Bureau of Standards defined:

1 yard = $\frac{3600}{3937}$ meter
= 0.9144 meter
1 inch = 25.4 millimeters
= 2.54 centimeters
1 U.S. pound = 0.453,592,427,7 kilogram
= 0.4536 (commonly used) kilogram
1 kilogram = 2.2046 pounds
1 U.S. gallon = 3,785.4 cubic centimeters
= 231 cubic inches
1 bushel = 2,150.42 cubic inches
= 0.0352 cubic meter
1 liter = 1 kilogram at 4°C.
= 1.0567 U.S. quarts
= 0.264 U.S. gallon

There is in Washington, D.C., a platinum-iridium graduated X cross section meter bar; and a platinum-iridium cylinder of equal diameter and height prototype kilogram, which are considered as the standards.

In 1960, the world adopted a new standard of length—a wave length of the orange red light from Krypton 86, which is 0.999996 U.S. unit. This new definition, so minutely changed, does not in any way change the relationship between U.S. and Metric units. It will be disregarded in this publication.

The British and the United States differ somewhat in the Weights and Measures and a separate discussion will be devoted to this.

I believe that machinery manufacturers who do considerable export business may be forced into the Metric System—except for the United States and the British Commonwealth, the entire remaining Free World utilizes the Metric System (Britain has recently announced that they will switch to the Metric System in 1975). However, it is doubtful indeed whether anything except feet, inches, rods, acres, etc., will ever be used in the United States for the millions of land deeds, mortgages, property titles, and architects and engineers building drawings.

HELMER A. RONNINGEN

EVERYDAY WEIGHTS AND MEASURES

U.S.

10 U.S. gallons of gasoline	=	37.85 liters
25 lbs./sq. in. auto tire pressure	=	1.76 kgs./cm.^{2}
2 lbs. candy	=	907.2 grams
16 ounce steak (1 lb.)	=	453.6 grams
1 inch	=	2.54 centimeters
1 foot = 12 inches	=	30.48 centimeters
1 mile = 5280 feet	=	1.609 kilometers

METRIC

10 liters of gasoline	=	2.65 gallons
1 kg./cm.^{2} auto tire pressure	=	14.22 lbs./sq. in.
500 grams candy	=	17.63 ounces
½ kilogram steak	=	1.102 pounds
1 centimeter (cm.)	=	0.3937 inch
1 meter (100 cms.)	=	39.37 inches
1 kilometer (1000 ms.)	=	0.621 mile (⅝ mile)

TONS

short or net ton	=	2,000 pounds (U.S.)
long or gross ton	=	2,240 pounds (British)
metric ton (1000 kgs.)	=	2,204.6 pounds (Metric)
ship register ton	=	100 cubic feet (ship tonnage)
measurement ton	=	40 cubic feet (cargo tonnage)
water ton	=	224 imperial gallons (British)
	=	2,240 pounds

NOTE:			
	1 lb.	=	0.4536 kg.
	1 cu. ft.	=	0.0283 cu. m.

U.S. TO METRIC

1 in.
2.54 cm.

1 in.
2.54 cm.

1 in.
2.54 cm.

1 sq. in.
= $(2.54)^2$ = 6.45 cm.^{2}

1 cu. in. = $(2.54)^3$
= 16.39 cm.^{3}

1 ft. = 12 in.
30.48 cm.

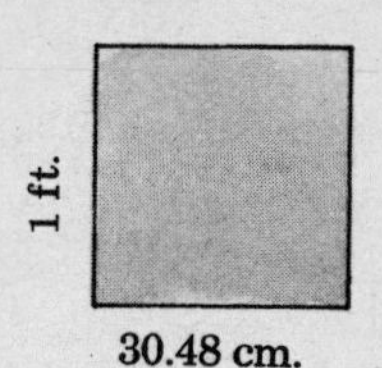

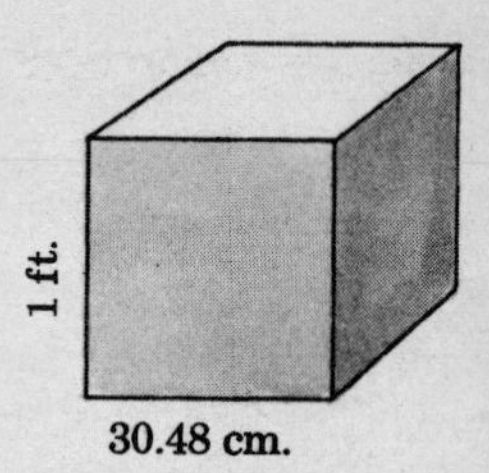

1 sq. ft. = 144 sq. in.
= $(30.48)^2$
= 929.03 cm.^{2}
= .093 m.^{2}

1 cu. ft. = 1728 cu. in.
= $(30.48)^3$
= 28,316.8 cm.^{3}
= .028 m.^{3}

METRIC TO U.S.

1 cm.
.394 in.

.394 in.
1 cm.

.394 in.
1 cm.

1 cm.^{2} = $(.394)^2$
= .155 sq. in.

1 cm.^{3} = $(.394)^3$
= .061 cu. in.

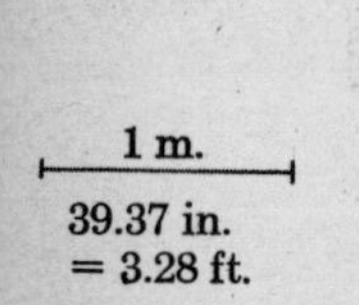

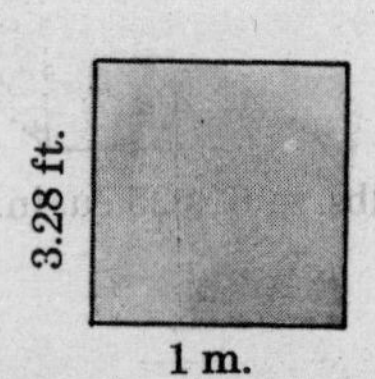

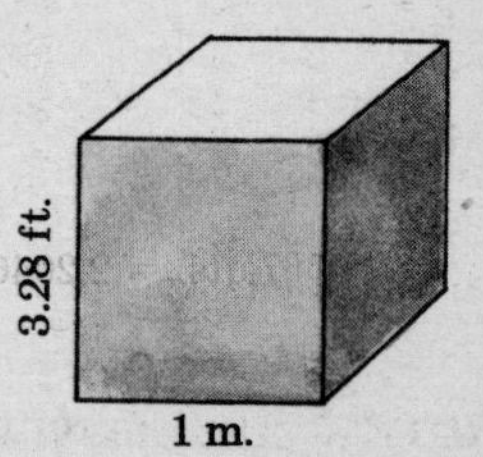

1 m.^{2} = $(3.28)^2$
= 10.76 sq. ft.

1 m.^{3} = $(3.28)^3$
= 35.31 cu. ft.

U.S. TO METRIC

1 gallon = 4 quarts = 8 pints = 8.345 lbs. = 231 cu. in. =

3.785 liters = 3.785 kg. = 3,785.4 cm.^{3}

METRIC TO U.S.

1 liter = 1 kg. = 1000 cu. cm. = cm.^{3} =

1.0567 qts. = 2.2046 lbs. = 61.025 cu. in.

NOTE: Above based on pure water
at maximum density at 4°C. = at 39.2°F.
1000 liters = 1000 kgs. = 1 m.^{3} = 1 metric ton
= 2204.6 lbs. = 35.31 cu. ft. = 264 gallons

METRIC PREFIXES

tera	=	10^{12}	=	1,000,000,000,000.
giga	=	10^{9}	=	1,000,000,000.
mega	=	10^{6}	=	1,000,000.
kilo	=	10^{3}	=	1,000.
hecto	=	10^{2}	=	100.
deka	=	10^{1}	=	10.
deci	=	10^{-1}	=	0.1
centi	=	10^{-2}	=	0.01
milli	=	10^{-3}	=	0.001
micro	=	10^{-6}	=	0.000,001
nano	=	10^{-9}	=	0.000,000,001
pico	=	10^{-12}	=	0.000,000,000,001

AREA

AREA

U.S. TO METRIC

1 square inch	=	6.4516 square centimeters
1 square foot	=	144 square inches
	=	0.093 square meter
1 square yard	=	9 square feet
	=	1,296 square inches
	=	0.836 square meter
1 square rod	=	30.25 square yards
	=	272.25 square feet
	=	25.293 square meters
1 acre	=	43,560 square feet
	=	160 square rods
	=	0.4047 hectare
1 square mile	=	640 acres
	=	259 hectares
	=	1 section of land
1 section of land	=	1 mile square
	=	259 hectares
1 township	=	6 miles square
	=	36 square miles
	=	36 sections
	=	23,040 acres
	=	9,324 hectares
	=	93.25 square kilometers
	=	9.66 kilometers square

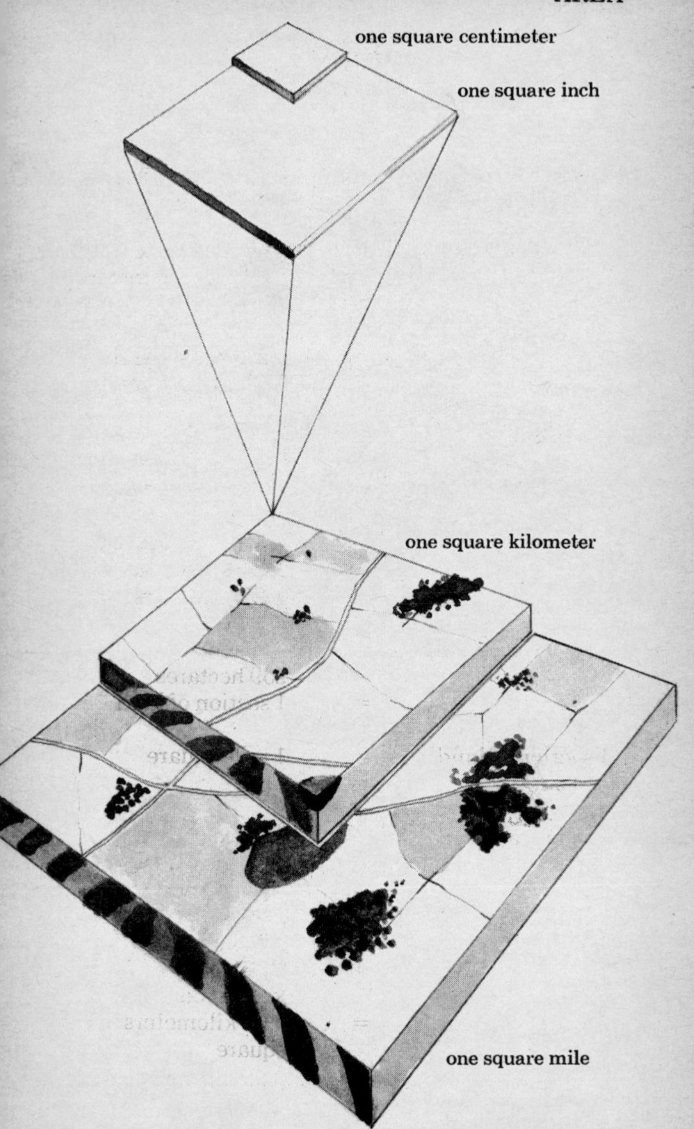
one square centimeter
one square inch
one square kilometer
one square mile

METRIC TO U.S.

1 square centimeter	=	100 square millimeters
	=	0.155 square inch
1 square meter	=	10,000 square centimeters
	=	10.764 square feet
	=	1.196 square yards
1 are	=	100 square meters
	=	0.0247 acre
1 hectare	=	100 ares
	=	10,000 square meters
	=	2.471 acres
1 square kilometer	=	1,000,000 square meters
	=	100 hectares
	=	247.11 acres
	=	0.386 square mile
100 square kilometers	=	10 kilometers square
	=	6.21 miles square
	=	38.56 square miles
	=	24,710.5 acres

SQUARE METERS
METRIC TO U.S.

10,000 square meters	=	1 hectare
	=	2.471 acres
100 square meters	=	1 are
	=	119.6 square yards
1 square meter	=	1 centare
	=	1550 square inches

AREA

One Square Inch = 6.4516 sq. cms.

One Square Centimeter = 0.155 sq. in.

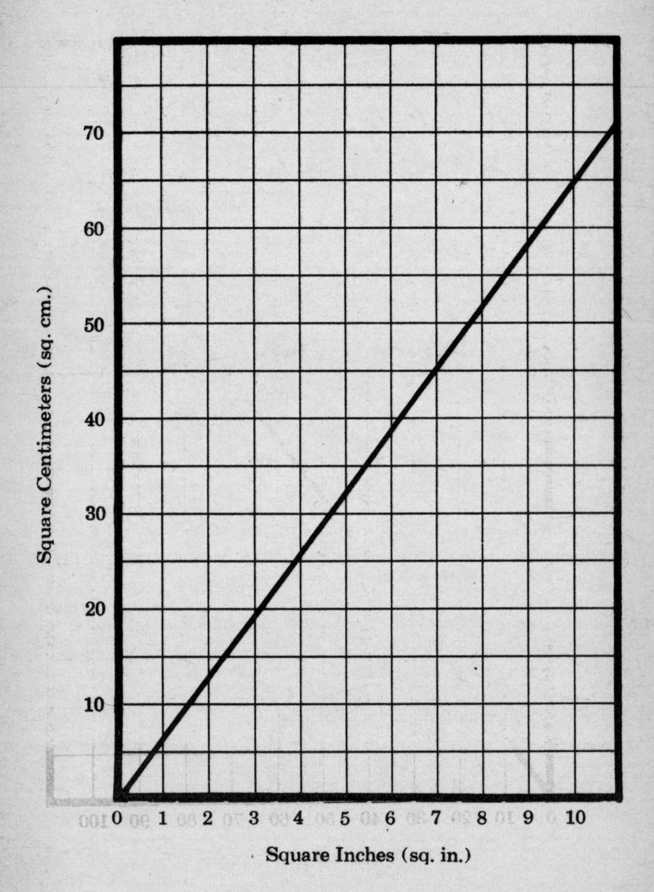

AREA

One Square Inch = 6.4516 sq. cms.

One Square Centimeter = 0.155 sq. in.

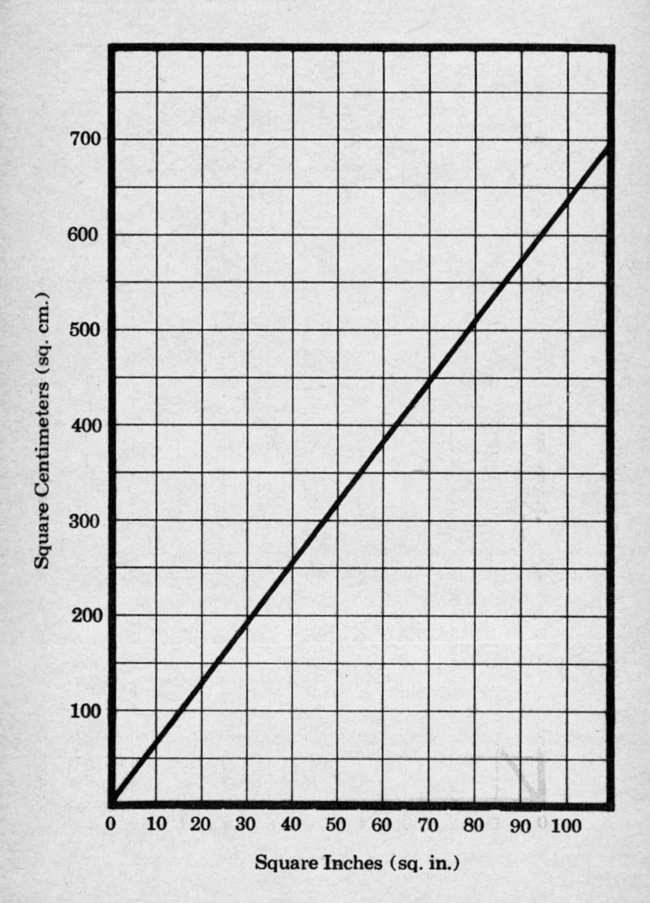

AREA

One Square Foot = 0.093 sq. m. or m.^{2}

One Square Meter = 10.763,91 sq. ft.
(Use = 10.76 sq. ft.)

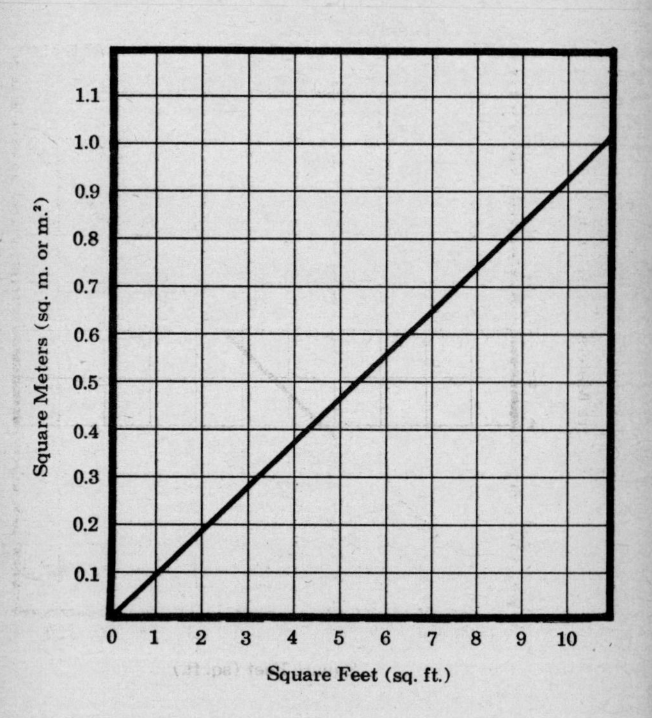

AREA

One Square Foot = 0.093 sq. m. or $m.^2$

One Square Meter = 10.763,91 sq. ft.
(Use = 10.76 sq. ft.)

One Square Yard = 9 sq. ft. = 0.836 sq. m.

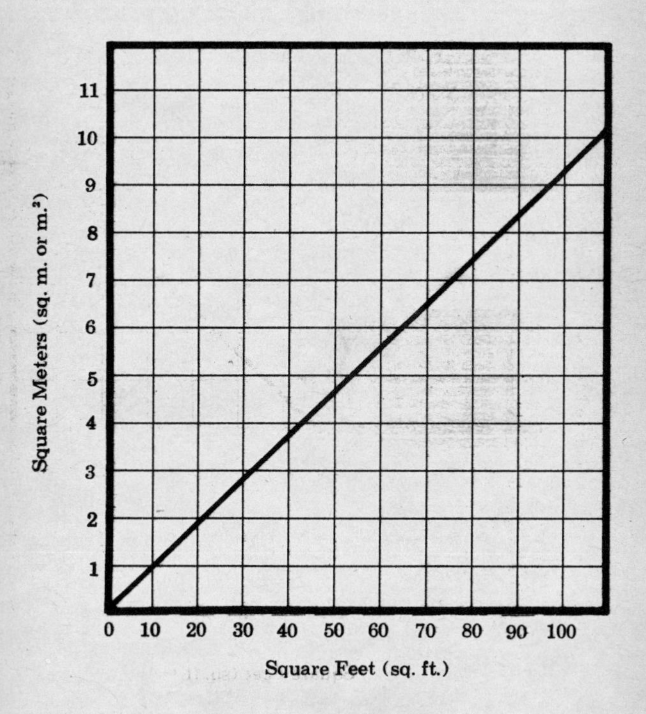

AREA

One Acre	=	43,560 sq. ft.
	=	0.404,685 Hectares
(Use	=	0.405 Hectares)

One Hectare	=	10,000 sq. m. ($m.^2$)
	=	2.471,05 Acres
(Use	=	2.47 Acres)

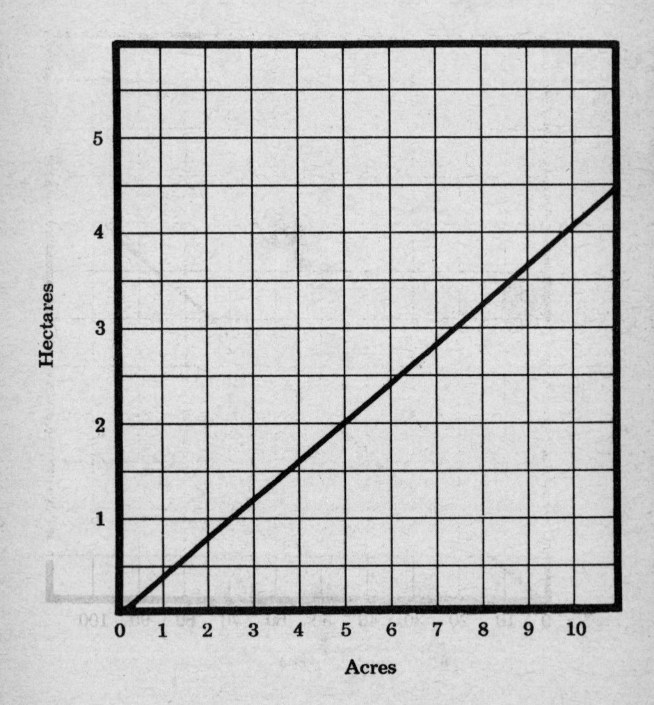

AREA

One Acre = 43,560 sq. m.
= 0.404,685 Hectares
(Use = 0.405 Hectares)

One Hectare = 10,000 sq. m. ($m.^2$)
= 2.471,05 Acres
(Use = 2.47 Acres)

640 Acres = 1 sq. m.
= 259 Hectares

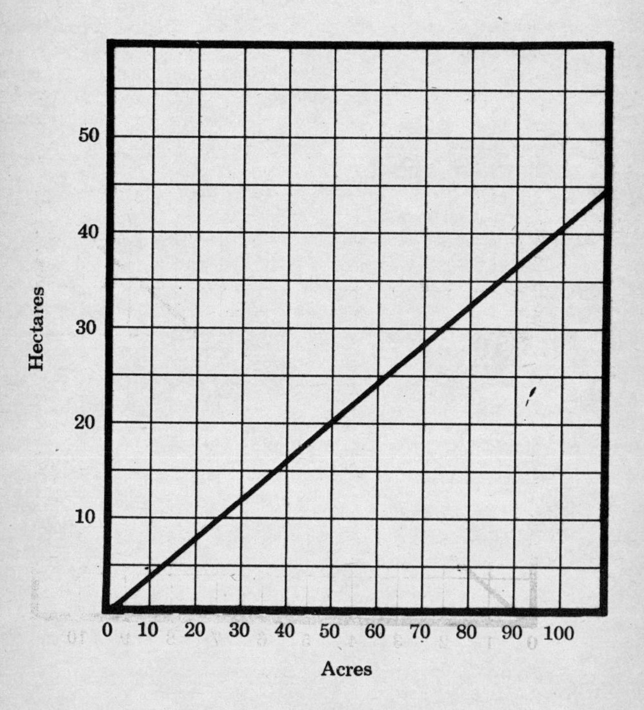

ENERGY, HEAT, POWER

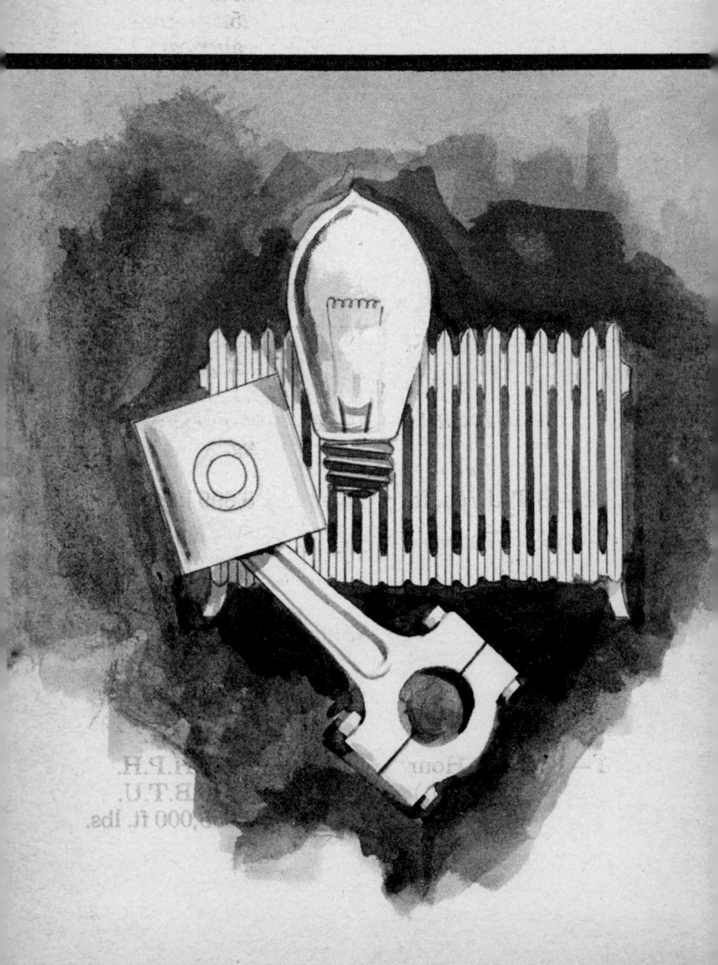

ENERGY, HEAT, POWER
U.S. TO METRIC

1 British Thermal Unit (B.T.U.)	=	heat required to raise 1 pound of water 1°F. at 39°F. – 4°C.
	=	252 gram calories
	=	778 foot pounds
1 foot pound	=	1 pound lifted vertically 1 foot
	=	0.3254 calorie
	=	1.356 joules
1 pound calorie	=	453.6 gram calories
1 – B.T.U./sq. ft.	=	0.2713 gr. cal./cm.^{2}
1 – B.T.U./lb.	=	0.5556 kg.-cal./kg.
1 – B.T.U./cu. ft.	=	8.899 kg.-cal./m.^{3}
1 – Horse Power Hour (H.P.H.)	=	33,000 foot lbs./minute
	=	2,545 B.T.U.
	=	1,980,000 ft. lbs.
	=	746 watts
1 – Kilowatt Hour (K.W.H.)	=	1.341 H.P.H.
	=	3,413 B.T.U.
	=	2,655,000 ft. lbs.

METRIC TO U.S.

joule	=	0.24 calorie
	=	0.7376 foot pounds
gram calorie	=	heat required to raise 1 gram of water 1°C. at 4°C.
	=	4.18 joules
	=	3.087 foot pounds
	=	0.00397 B.T.U.
kilogram calorie	=	heat required to raise 1,000 grams or 1 kilogram or 1 liter of water 1°C. at 4°C.–39°F.
	=	3.969 B.T.U.
gr.-cal./cm.^{2}	=	3.687 B.T.U. /sq. ft.
kg.-cal./m.^{3}	=	0.1124 B.T.U./cu. ft.
Kilowatt Hour (K.W.H.)	=	2,655,000 foot pounds
	=	3,413 B.T.U.
	=	1.341 H.P.H.

LENGTH

LENGTH

COMMON SYMBOLS

inch	=	in. (″)
feet	=	ft. (′)
yard	=	yd.
kilometer	=	km.
meter	=	m.
centimeter	=	cm.
millimeter	=	mm.
micron	=	μ

LINEAR MEASUREMENT
U.S. TO METRIC

1 inch	=	2.54 centimeters (cms.)
	=	25.4 millimeters
	=	25,400 microns
1 foot	=	12 inches
	=	30.48 centimeters
	=	0.3048 meter
1 yard	=	3 feet
	=	36 inches
	=	91.44 centimeters
	=	0.9144 meter
1 fathom	=	6 feet
	=	2 yards
	=	1.8288 meters
1 rod	=	16.5 feet
	=	5.5 yards
	=	5.029 meters
1 furlong	=	40 rods
	=	660 feet
	=	220 yards
	=	⅛ mile
	=	201.168 meters
1 mile	=	8 furlongs
	=	5,280 feet
	=	1,760 yards
	=	1,609.344 meters

METRICS

mile/myriameter

mile/kilometer

mile/hectometer

yard/dekameter

yard/meter

inch/decimeter

inch/centimeter

inch/millimeter

inch/micron

METRIC TO U.S.

myriameter	=	10,000 meters
	=	10 kilometers
	=	6.21 miles
kilometer	=	1,000 meters
	=	10 hectometers
	=	0.621,37 mile
	=	3280.8 feet
hectometer	=	100 meters
	=	10 dekameters
	=	0.062 mile
	=	328.08 feet
dekameter	=	10 meters
	=	32.808 feet
meter	=	10 decimeters
	=	100 centimeters
	=	1,000 millimeters
	=	3.2808 feet
	=	39.37 inches
decimeter	=	10 centimeters
	=	100 millimeters
	=	3.937 inches
centimeter	=	10 millimeters
	=	0.3937 inch
millimeter	=	1,000 microns
	=	0.039,37 inch
micron	=	.001 millimeter
	=	.000,039,4 inch
	=	.000,04 inch (normally used)

SURVEYORS CHAIN MEASUREMENT

1 link	=	7.92 inches
	=	20.11 centimeters
100 links	=	1 chain
	=	4 rods
	=	66 feet
	=	20.117 meters
80 chains	=	1 mile
	=	320 rods
	=	5,280 feet
	=	1,760 yards
	=	1,609.33 meters

NAUTICAL MEASUREMENT

1 nautical mile	=	6,076.11 feet
	=	1,852 meters

SCANDINAVIAN MILE

1 scandinavian mile	=	10 kilometers
	=	6.2137 U.S. miles

POINT (TYPOGRAPHY)

1 point	=	0.013,84 inch
	=	1/72 inch (approx.)
	=	0.351 millimeters

METERS
METRIC TO U.S.

10,000 meters	=	1 myriameter
	=	6.2137 miles
1,000 meters	=	1 kilometer
	=	0.621,37 mile
100 meters	=	1 hectometer
	=	328 feet and 1 inch
10 meters	=	1 dekameter
	=	393.7 inches
1 meter	=	39.37 inches
1/10 of a meter	=	1 decimeter
	=	3.937 inches
1/100 of a meter	=	1 centimeter
	=	0.3937 inch
1/1000 of a meter	=	1 millimeter
	=	0.0394 inch
1/1,000,000 of a meter	=	1 micron
	=	0.000,039,4 inch
	=	0.000,04 inch (normal)

LENGTH

One Micron = 0.001 mm.
= 0.000,039,37 in.
= 0.000,04 in.
(equivalent use.)

One Inch = 25.4 mm.
= 25,400 microns

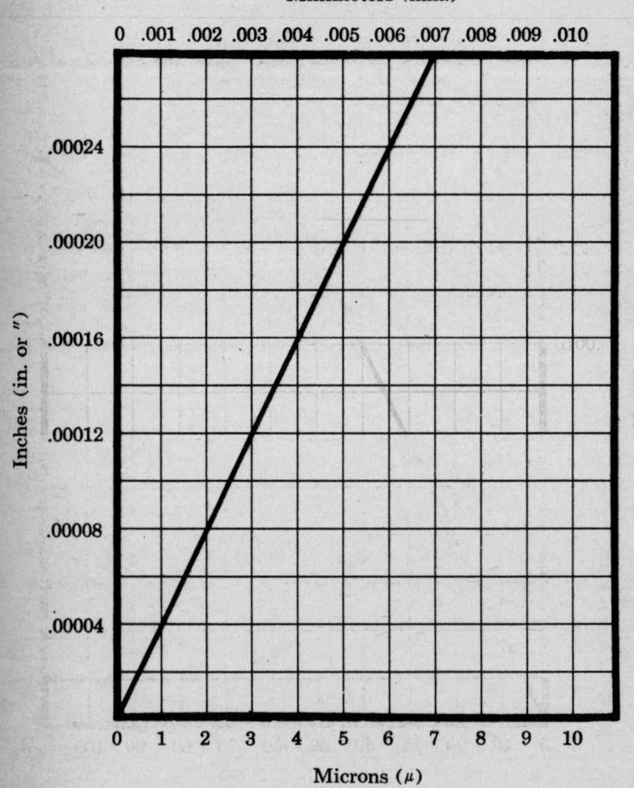

LENGTH

One Micron = 0.001 mm.
= 0.000,039,37 in.
= 0.000,04 in.
(equivalent use.)

One Inch = 25.4 mm.
= 25,400 microns

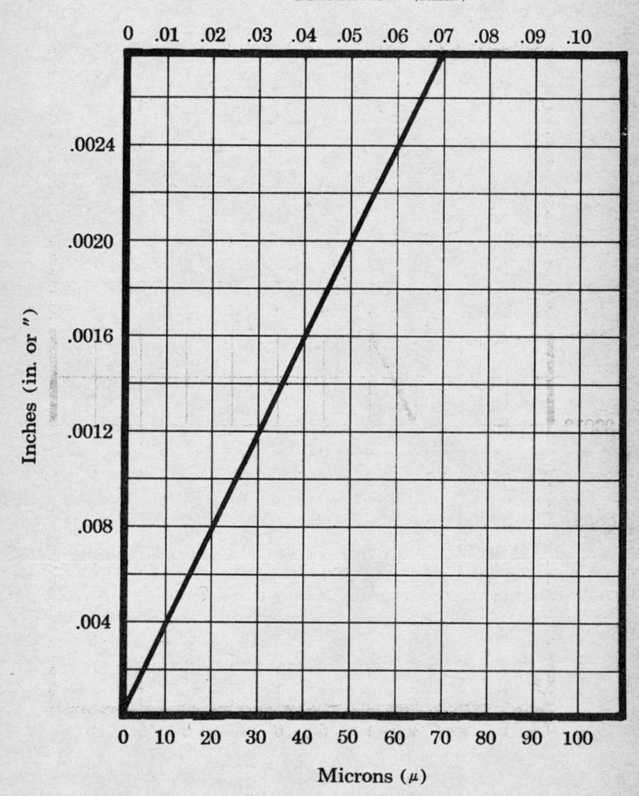

LENGTH

One Inch = 2.54 cms.

One Centimeter = .3937 in.

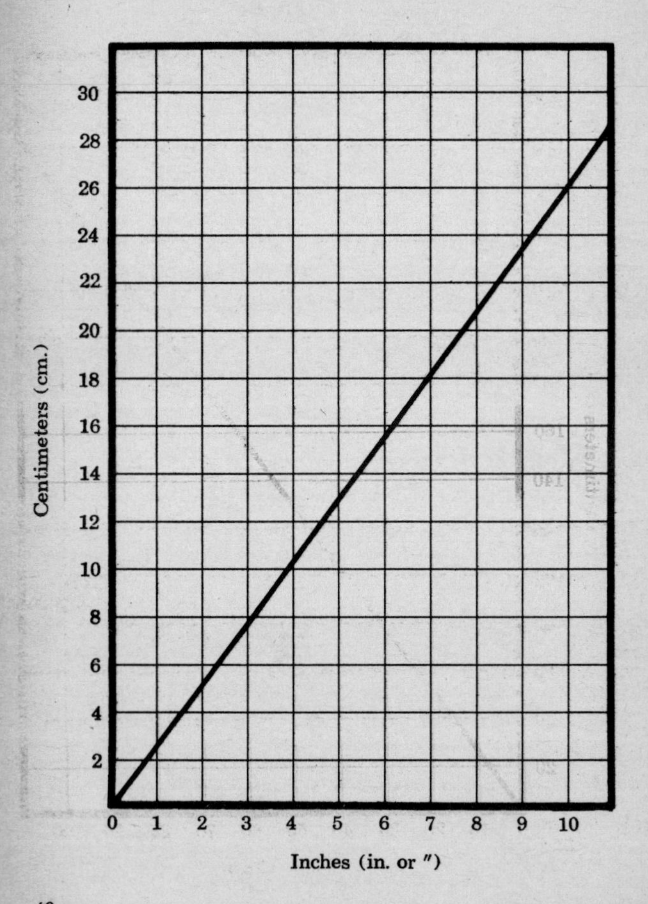

LENGTH

One Inch = 2.54 cms.

One Centimeter = .3937 in.

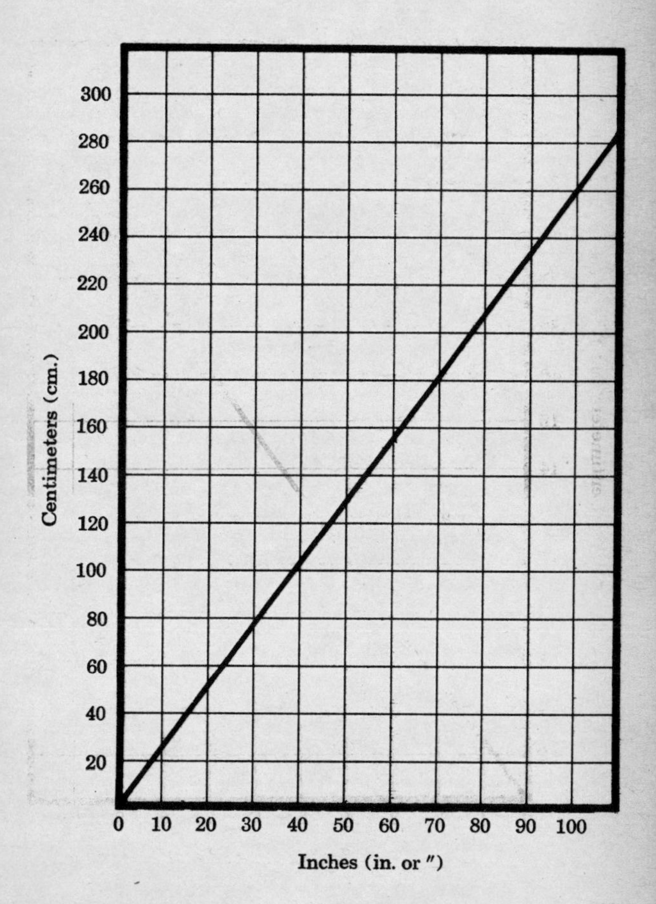

LENGTH

One Foot = 12 inches
= 0.3048 meter
= 30.48 centimeters

One Meter = 100 centimeters
= 39.37 inches
= 3.28 feet

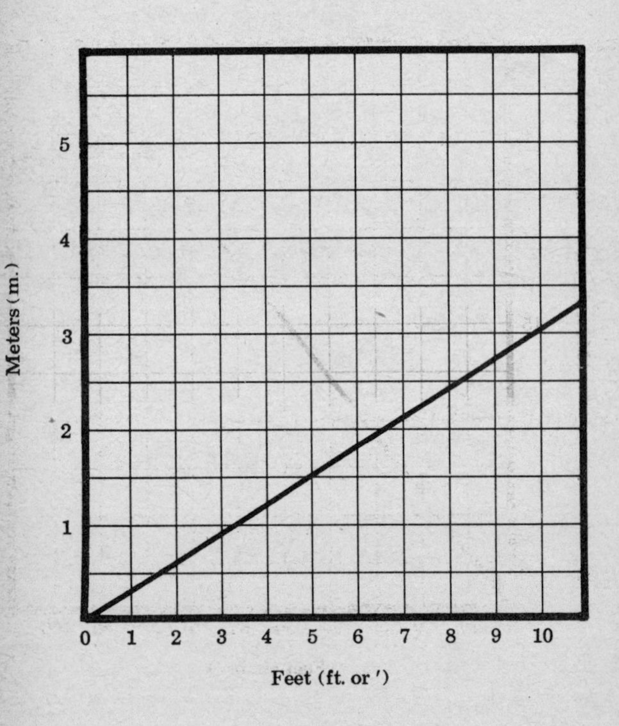

LENGTH

One Foot = 12 inches
= 0.3048 meter
= 30.48 centimeters

One Meter = 100 centimeters
= 39.37 inches
= 3.28 feet

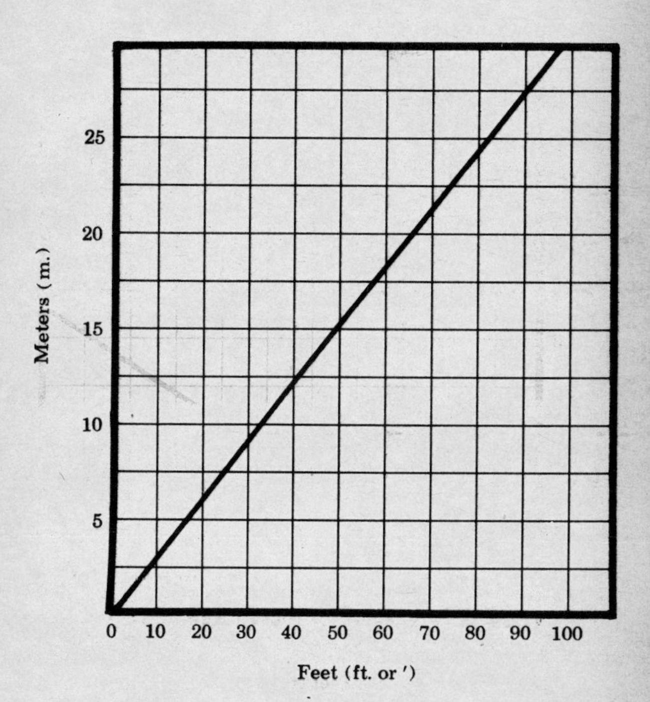

LENGTH

One U.S. Mile = 5,280 feet
= 1.609 kilometers

One Kilometer = 1,000 meters
= .621 mile
= 3279 feet

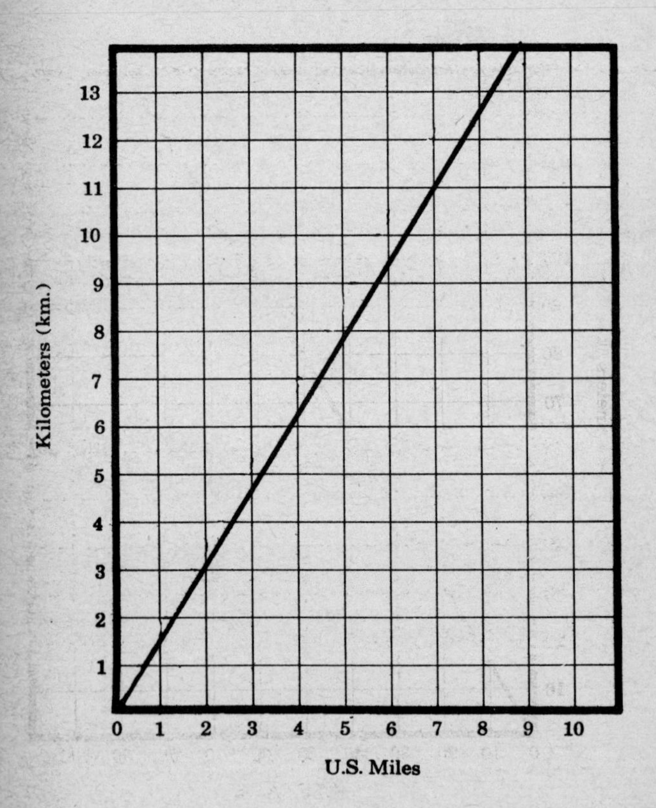

LENGTH

One U.S. Mile = 5,280 feet
= 1.609 kilometers

One Kilometer = 1,000 meters
= .621 mile
= 3279 feet

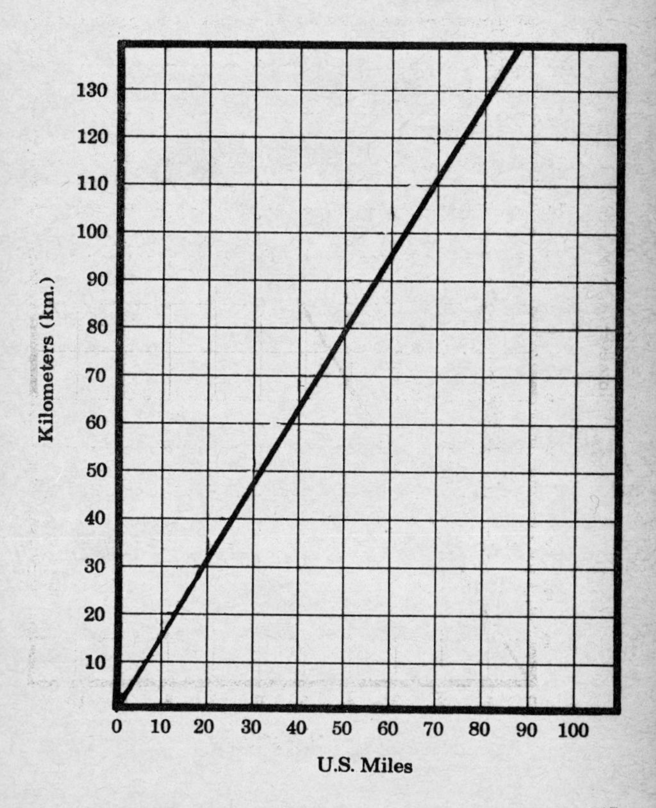

TEMPERATURE

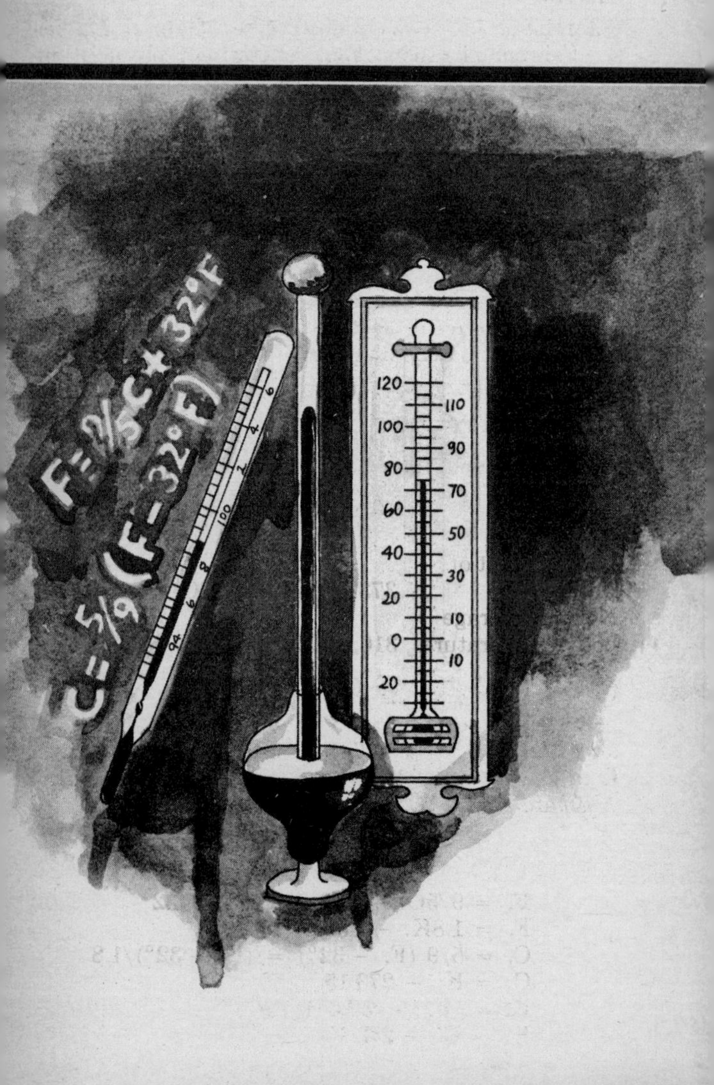

TEMPERATURE

Fahrenheit — F.: Centigrade — C. — Celsius: Kelvin — K.

Fahrenheit—F.—was named after Gabriel Daniel Fahrenheit (1686–1736), a German physicist in which he defined: boiling point of water—212°F.; freezing point of water—32°F.;lowest temperature obtained by mixing equal parts of snow and salt —0°F. (all at sea level)

Note: Boiling point of water is reduced approximately 2% for every 1,000 feet above sea level: at 5,000 feet this would be 212°F. −10°F. =202°F.

Celsius—C.— Commonly known as Centigrade— see next page

Kelvin—K.— Named after British Lord William Thomson Kelvin (1824–1907) who developed scale of absolute zero—0°A.—which is the hypothetical point in which a substance would have no heat, and no molecular motion.

	Kelvin (K.)	Celsius (C.)	Fahrenheit (F.)
Boiling point of water	373.15°K. =	100°C. =	212°F.
Normal average body temperature	310.15°K. =	37.0°C. =	98.6°F.
Water freezes	273.15°K. =	0°C. =	32.0°F.
Lowest temperature obtained by mixing equal parts snow and salt	255.37°K. =	−17.8°C. =	0°F.
C and F the same	233.15°K. =	−40°C. =	−40°F.
Absolute zero	0°K. =	−273.15°C. =	−459.67°F.

CONVERSIONS

F. = 9/5C. + 32° = 1.8C. + 32°
F. = 1.8K. − 459.67
C. = 5/9 (F. − 32°) = (F. − 32°)/1.8
C. = K. − 273.15
K. = (F. + 459.67)/1.8
K. = C. + 273.15

CELSIUS–CENTIGRADE THERMOMETER –C.°

Centigrade Thermometer was named after Anders Celsius (1701-1744) a Swedish astronomer and scientist, Upsala Sweden. He proposed:

Boiling Point of Water	100°C.	=	212°F.
Freezing Point of Water	0°C.	=	32°F.

All at sea level barometric pressure.

METRIC TO U.S.

100°C.	= 212°F.	=	Boiling Point of Water
60°C.	= 140°F.	=	Hot Water
40°C.	= 104°F.	=	Warm Water
25°C.	= 77°F.	=	Warm Room Temperature
7°C.	= 45°F.	=	Cold Water
0°C.	= 32°F.	=	Freezing Point
−17.8°C.	= 0°F.	=	Coldest Point of Water, Equal mixture snow and salt
−40°C.	= −40°F.		

C. = 5/9 (F. − 32°)
F. = 9/5 C. + 32°F.

Many scientists and engineers are using Kelvin–K.– Absolute Zero in their research work as an international standard (SI).

212°F.	= 100°C.	= 373.15°K.	
32°F.	= 0°C.	= 273.15°K.	
—459.67°F.	= —273.15°C.	= 0°K.	= Absolute Zero

20°C

20°F

TEMPERATURE

Centigrade (C.)   Fahrenheit (F.)

$$C. = 5/9\,(F. - 32°)$$

$$F. = 9/5\,C. + 32°F.$$

Boiling Point of Water = 100°C. = 212°F.

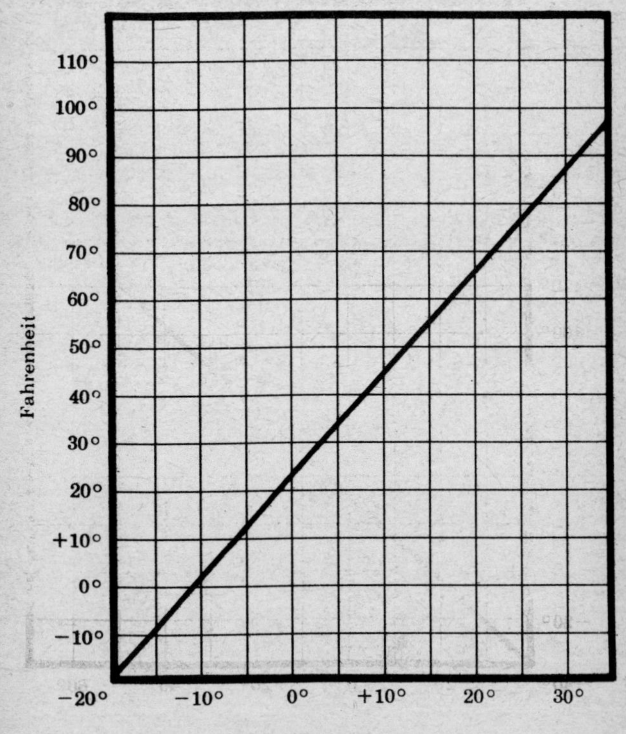

TEMPERATURE

Centigrade (C.) Fahrenheit (F.)

$$C. = 5/9\,(F. - 32^\circ)$$

$$F. = 9/5\,C. + 32^\circ F.$$

Freezing Point of Water = 0°C. = 32°F.

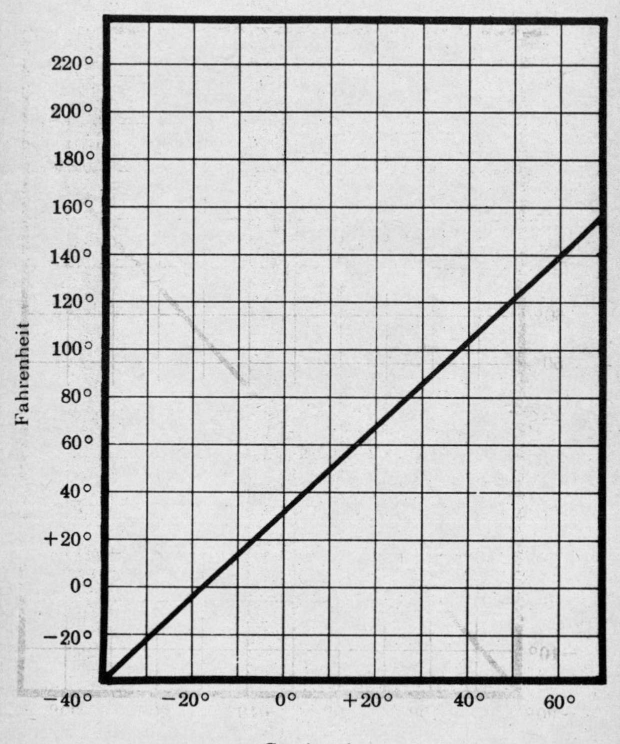

PRESSURE

PRESSURE
U.S. TO METRIC

pounds per square inch	=	lbs./sq. in.
lbs./sq. in.	=	144 lbs./sq. ft.
	=	70.3 gr./cm.^{2}
	=	0.0703 kgs./cm.^{2}
	=	703 kgs./m.^{2}
	=	0.068 atmospheres
	=	2.31 feet of water
	=	2.04 inches of mercury
	=	51.71 mm. of mercury
lbs./sq. ft.	=	478.8 dynes per sq. cm.
	=	4.882 kgs./sq. m.
	=	0.488 gram/sq. cm.
	=	0.016 ft. of water
	=	0.359 mm. of mercury at 32°F. at 0°C.
inch of water at 4°C.	=	2,490.82 dynes/sq. cm.
	=	25.4 kg./m.^{2}
	=	5.2 #/sq. ft.
	=	2.54 gr./cm.^{2}
	=	1.86 mm. of mercury
foot of water at 4°C.	=	304.79 kg./sq. cm.
	=	62.426 #/sq. ft.
	=	30.48 gram/cm.^{2}
	=	22.42 mm. of mercury at 4°C.
inch of mercury	=	70.73 #/sq. ft.
	=	0.491 #/sq. in.
	=	33,863.9 dynes/cm.^{2}
	=	345.3 kgs./m.^{2}
	=	34.53 grams/cm.^{2}

METRIC TO U.S.

1 gram/cm.^{2}	=	10 kg./m.^{2}
	=	980.7 dynes/cm.^{2}
	=	2.048 lbs./sq. ft.
1 kgs./cm.^{2}	=	980,665 dyne/cm.^{2}
	=	2,048.2 lbs./sq. ft.
	=	14.22 lbs./sq. in.
	=	735.6 mm. of mercury at 4°C.
	=	32.81 ft. of water
	=	0.97 atmospheres
1 kg./m.^{2}	=	0.205#/sq. ft.
	=	98.07 dynes/cm.^{2}
	=	0.1 gram/cm.^{2}
	=	0.0736 mm. of mercury at 4°C.
	=	0.0393 in. of water at 4°C.
1 metric ton/m.^{2}	=	204.82 lbs./sq. ft.
	=	1,000 kgs./m.^{2}
	=	100 grams/cm.^{2}
	=	3.281 ft. of water at 4°C.
	=	73.56 mm. of mercury at 4°C.

NOTE:

U.S. auto tire pressures of:

25 lbs/sq. in = 1.757 kg./cm.^{2}

30 lbs/sq.in = 2.109 kg./cm.^{2}

1Kgs/cm²

Pressure

32.81
FT.

PRESSURE

One Pound Per Square Inch (lb./sq. in.)

lb./sq. in. = 0.0703 kgs./cm.²
= 0.068 atmospheres

kg./cm.² = 14.22 lbs./sq. in.
= 0.97 atmospheres

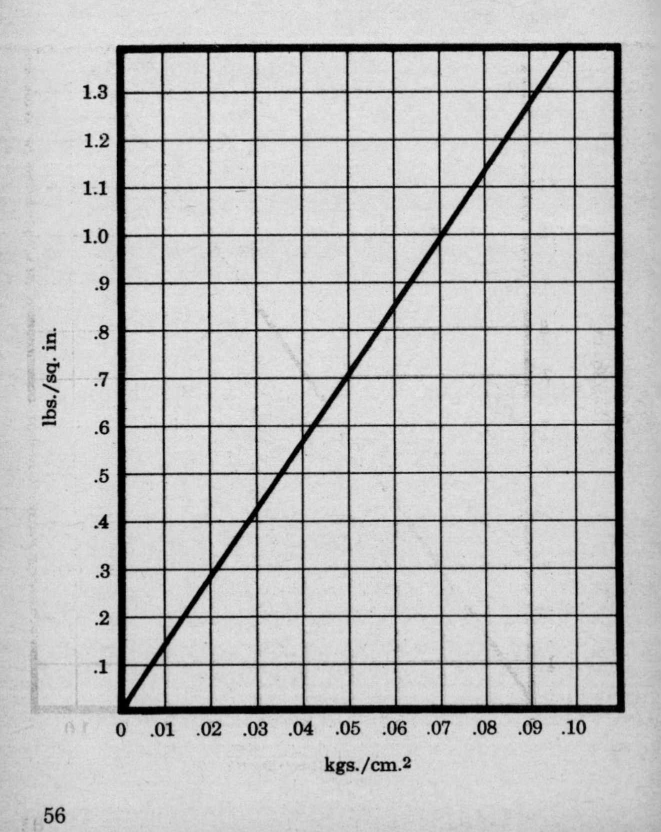

PRESSURE

lb./sq. in. = 0.0703 kg./cm.^{2}
= 0.068 atmospheres

kg./cm.^{2} = 14.22 lbs./sq. in.
= 0.97 atmospheres

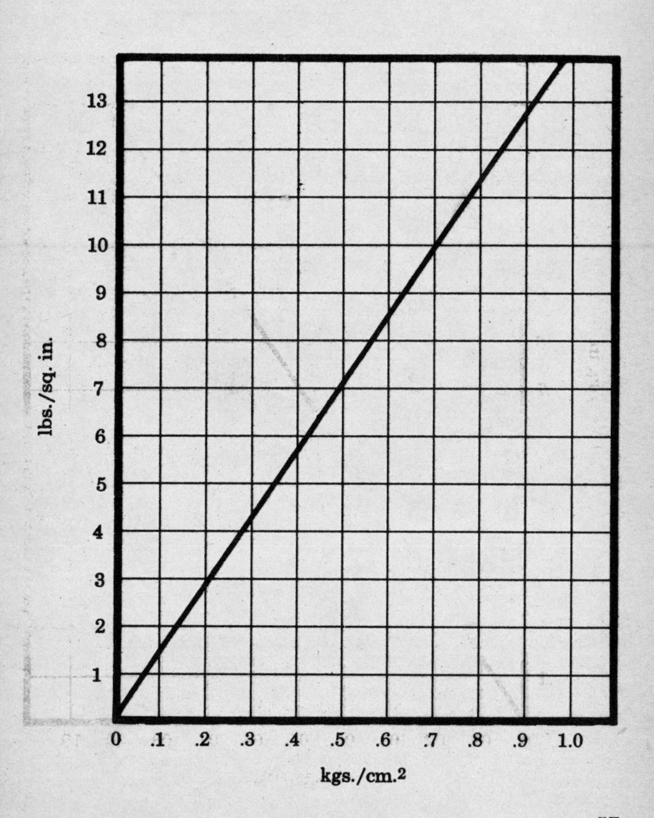

PRESSURE

lb./sq. in. = 0.0703 kgs./cm.²
= 0.068 atmospheres

kg./cm.² = 14.22 lbs./sq. in.
= 0.97 atmospheres

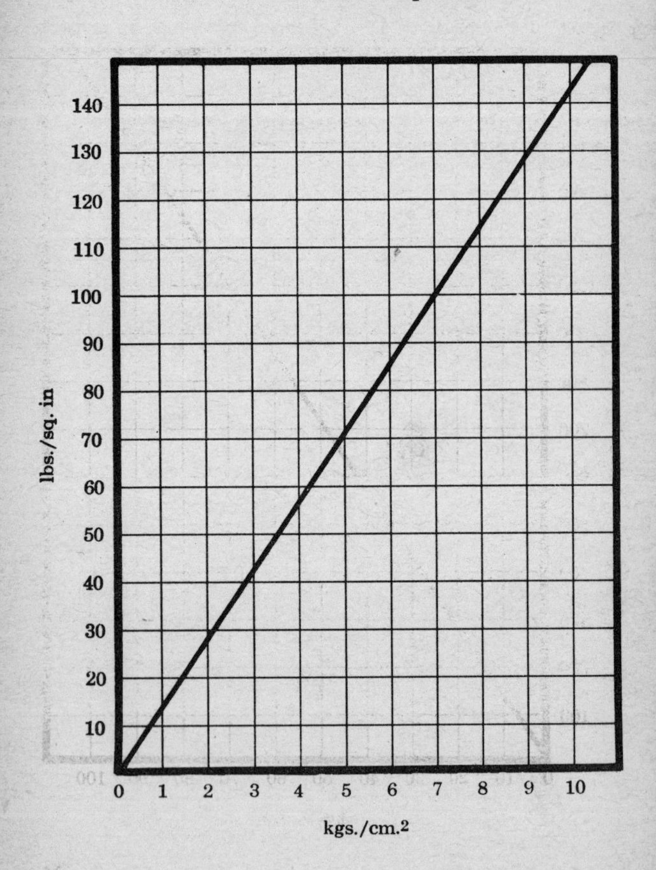

PRESSURE

lb./sq. in. = 0.0703 kgs./cm.^{2}
= 0.068 atmospheres

kg./cm.^{2} = 14.22 lbs./sq. in.
= 0.97 atmospheres

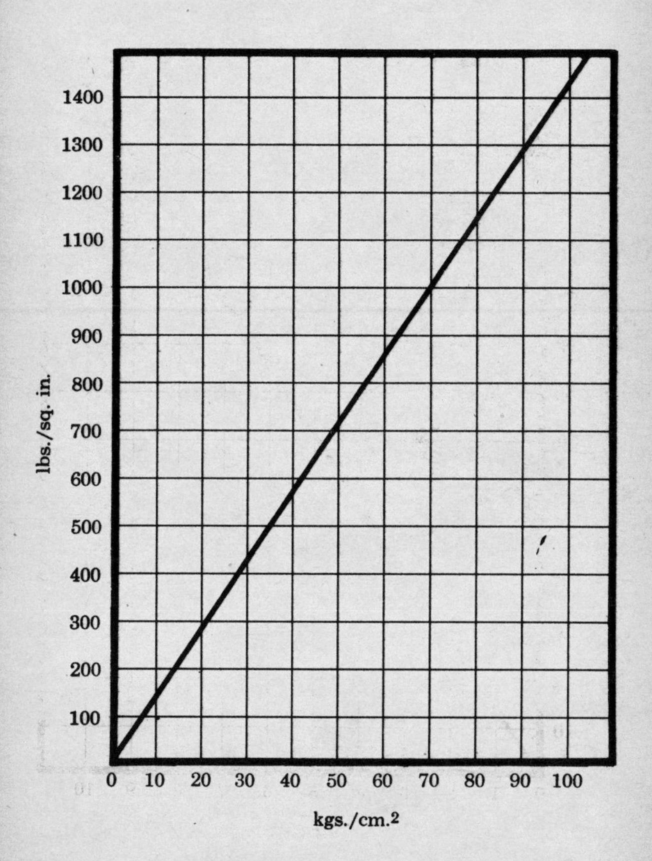

VOLUME

ONE HALF FOOT
TWO FEET
CUBIC FEET
208.17 LITERS
5 GAL.
3.785 LITERS

VOLUME
LIQUID MISCELLANEOUS

1 liter of water at 4°C. (39.2°F.) weighs 1 kilogram	=	2.2046 U.S. pounds
	=	1.057 quarts
	=	0.264 gallon
1 gallon of water	=	3.785 liters
	=	8.345 pounds
	=	4 quarts
1 cubic foot of water	=	7.481 gallons
	=	28.316 liters
	=	62.427 pounds
1 barrel of water	=	55 U.S. gallons
	=	208.17 liters
1 barrel of petroleum	=	42 gallons
	=	158.97 liters
1 barrel of beer	=	31 gallons
	=	117.33 liters

LIQUID CAPACITY
METRIC TO U.S.

1 minim	=	0.017 dram
	=	0.002 fluid ounce
1 dram	=	60 minims
	=	0.125 fluid ounce
	=	0.031 gill
1 ounce	=	480 minims
	=	8 fluid drams
	=	0.25 gill
1 gill	=	1,920 minims
	=	32 fluid drams
	=	4 fluid ounces
1 pint	=	16 fluid ounces
	=	4 gills
1 liquid quart	=	256 fluid drams
	=	32 fluid ounces
1 liquid gallon	=	1,024 drams
	=	128 ounces
	=	8 pints
	=	4 quarts
	=	231 cubic inches
	=	3.785 liters
	=	0.1337 cubic feet
1 cubic inch	=	4.433 drams
	=	0.554 fluid ounces
	=	0.034 pints
	=	16.386 milliliters
	=	0.016 liters

METRIC TO U.S.

1 kiloliter	=	1,000 liters
	=	10 hectoliters
	=	264 gallons
1 hectoliter	=	100 liters
	=	10 dekaliters
	=	26.4 gallons
1 dekaliter	=	10 liters
	=	2.64 gallons
	=	10.56 quarts
1 liter	=	1,000 milliliters
	=	10 deciliters
	=	0.264 gallon
	=	1.056 quarts
	=	61.025 cubic inches
1 deciliter	=	100 milliliters
	=	10 centiliters
	=	0.1056 quart
1 centiliter	=	10 milliliters
	=	0.010,56 quart
1 milliliter	=	.001 liter
	=	0.001,056 quart

METRIC TO U.S.

1,000 liters	=	1 kiloliter
	=	1 cubic meter
	=	1.308 cubic yards
	=	264.17 gallons
100 liters	=	1 hectoliter
	=	1/10 of a cubic meter
	=	26.4 gallons, liquid
10 liters	=	1 dekaliter
	=	10 cubic decimeters
	=	9.08 quarts, dry
	=	2.64 gallons, liquid
1 liter	=	1 cubic decimeter
	=	1 kilogram
	=	1.0567 quarts, liquid
1/10 of a liter	=	6.1022 cubic inches, dry
	=	0.845 gill, liquid
1/100 of a liter	=	1 centiliter
	=	10 cubic centimeters
	=	0.6102 cubic inch, dry
	=	0.338 fluid ounce
1/1000 of a liter	=	1 milliliter
	=	1 cubic centimeter
	=	0.061 cubic inch, dry
	=	0.27 fluid dram

VOLUME

Liquid, Capacity

One U.S. Gallon = 3.785 liters

One Liter = 0.2642 U.S. gallon

NOTE: 1 U.S. gal. = .833 imperial gal.

1 imperial gal. = 1.2 U.S. gal.

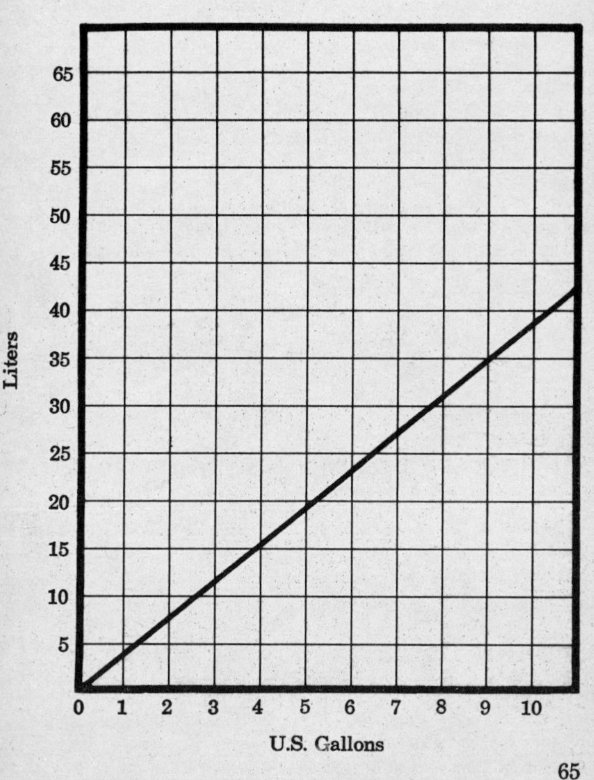

VOLUME

Liquid, Capacity

One U.S. Gallon = 3.785 liters

One Liter = 0.2642 gallon

NOTE: 1 U.S. gal. = .833 imperial gal.

1 imperial gal. = 1.2 U.S. gal.

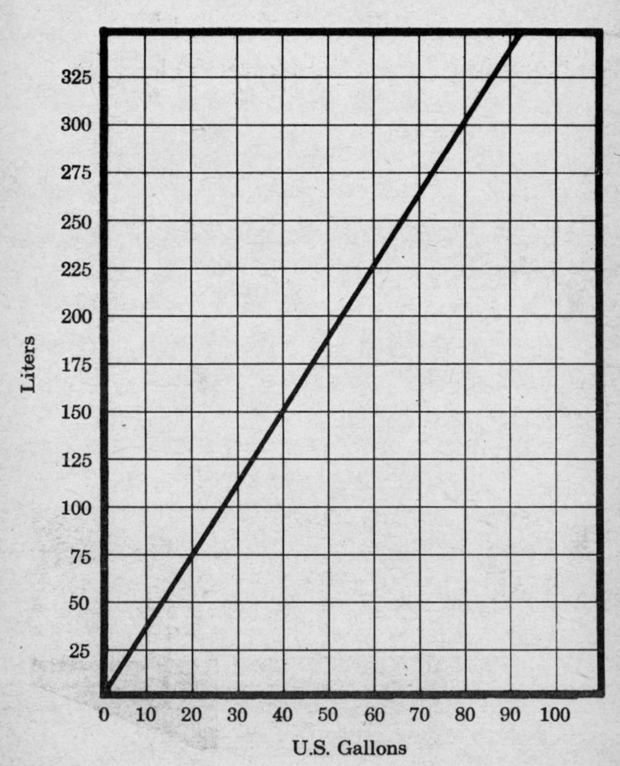

DRY, AIR OR GAS
U.S. TO METRIC

1 cubic inch	=	16.387 cubic centimeters
1 cubic foot	=	1,728 cubic inches
	=	28.317 cubic decimeters
	=	0.0283 cubic meter ($m.^3$)
1 cubic yard	=	27 cubic feet
	=	764.554 cubic decimeters
	=	0.7646 cubic meter ($m.^3$)
1 acre foot	=	1,233.48 cubic meters
	=	43,560 cubic feet

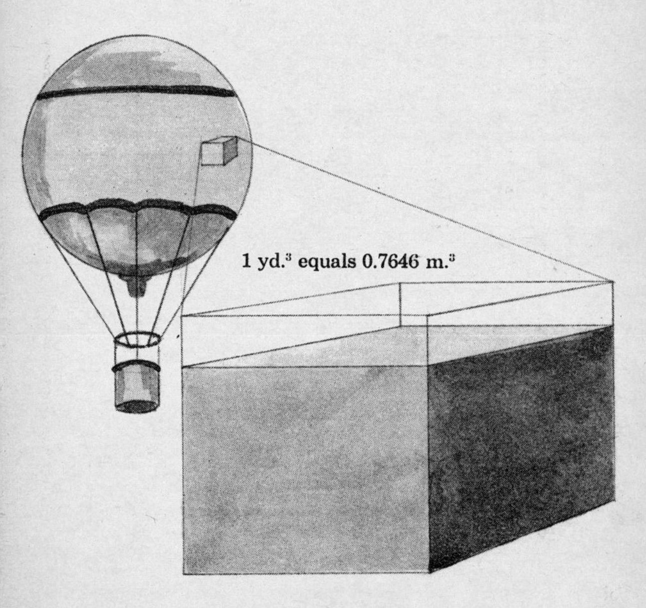

METRIC TO U.S.

1 cubic centimeter	=	1,000 cubic millimeters
	=	0.061 cubic inch
	=	0.001 cubic decimeter
1 cubic decimeter	=	1,000 cubic centimeters
	=	0.001 cubic meter
	=	61.0237 cubic inches
	=	0.0353 cubic foot
1 cubic meter ($m.^3$)	=	1,000 cubic decimeters
	=	35.3147 cubic feet
	=	1.3079 cubic yards
1,000 cubic meters	=	0.8107 acre feet

DRY, SOLIDS
U.S. TO METRIC

1 dry pint	=	33.6 cubic inches
	=	0.551 liter
1 dry quart	=	2 pints
	=	67.20 cubic inches
	=	1.101 liters
1 dry peck	=	8 quarts
	=	16 pints
	=	537.605 cubic inches
	=	8.809 liters
1 bushel	=	4 pecks
	=	32 quarts
	=	2,150.42 cubic inches
	=	1.244 cubic feet
	=	35.238 liters
1 cubic foot	=	25.71 dry quarts
	=	0.804 bushel
	=	1,728 cubic inches
	=	28.316 liters
	=	0.0283 cubic meter
1 cubic yard	=	27 cubic feet
	=	764.53 liters
	=	0.765 cubic meter
1 cord of wood	=	128 cubic feet
	=	4 ft. x 4 ft. x 8 ft.
	=	3.625 cubic meters (m.^{3})

VOLUME

Dry: Air, Gas, Solids

One Cubic Foot = 0.028,32 cu. m.

One Cubic Meter (m.³) = 35.31 cu. ft.

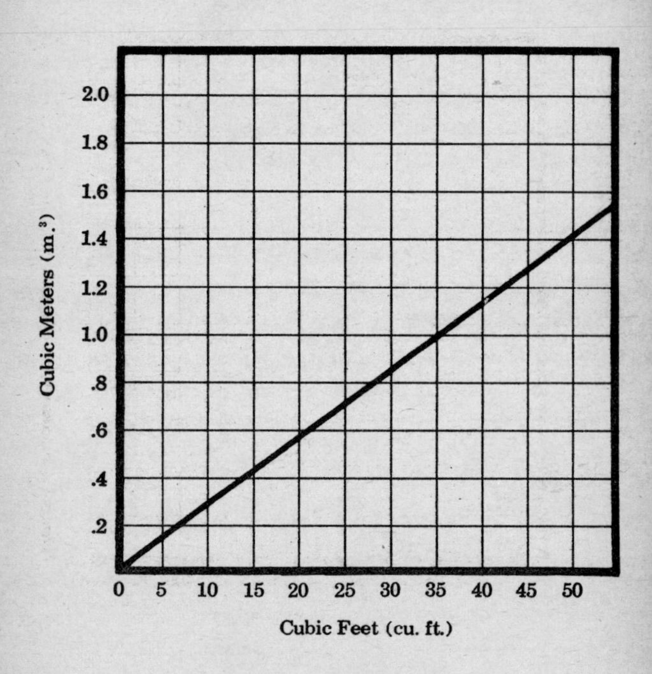

VOLUME

Dry: Air, Gas, Solids

One Cubic Foot = 0.028,32 cu. m.

One Cubic Meter ($m.^3$) = 35.31 cu. ft.

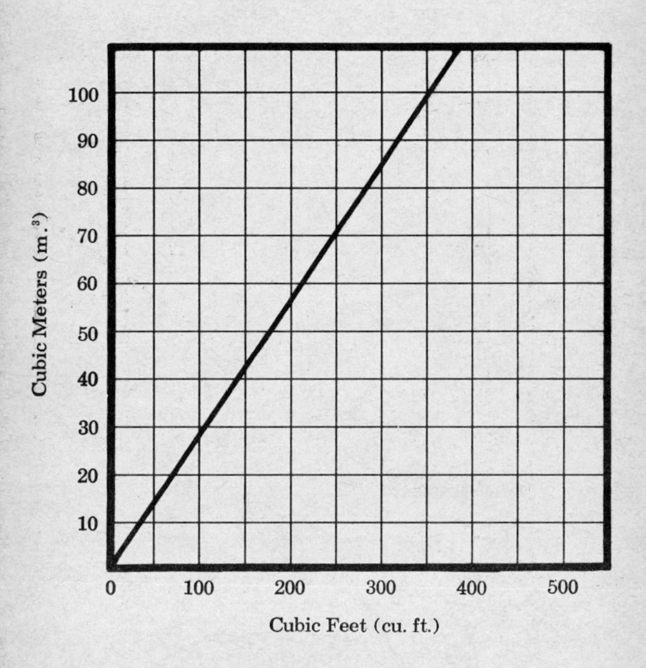

BRITISH
IMPERIAL DRY OR LIQUID CAPACITY TO METRIC

1 imperial gill	=	5 ounces
	=	141.75 grams
1 imperial pint	=	4 gills
	=	20 ounces
	=	567 grams
	=	0.567 liter
1 imperial quart	=	2 pints
	=	40 ounces
	=	1.136 kilograms
	=	1.136 liters
1 imperial gallon	=	4 quarts
	=	160 ounces
	=	10 pounds
	=	1.20 U.S. gallons
	=	4.546 kilograms
	=	4.546 liters
1 imperial peck	=	2 gallons
	=	20 pounds
	=	9.072 kilograms
	=	9.072 liters
1 imperial bushel	=	8 gallons
	=	80 pounds
	=	36.288 kilograms
1 imperial quarter	=	8 bushels
	=	290.304 kilograms
	=	290.304 liters

NOTE: British pound = U.S. pounds
British gallon = 1.20 U.S. gallons

METRIC TO IMPERIAL DRY OR LIQUID CAPACITY

100 grams	=	0.1 liter
	=	0.1 kilogram
	=	0.022 imperial gallon
	=	0.22 pound
	=	3.52 ounces
1 liter	=	1 cubic decimeter
	=	1 kilogram
	=	0.22 imperial gallon
	=	2.2 pounds
	=	35.2 ounces
1 cubic meter	=	1,000 kilograms
	=	1,000 cubic decimeters
	=	219.97 imperial gallons
	=	220.46 pounds

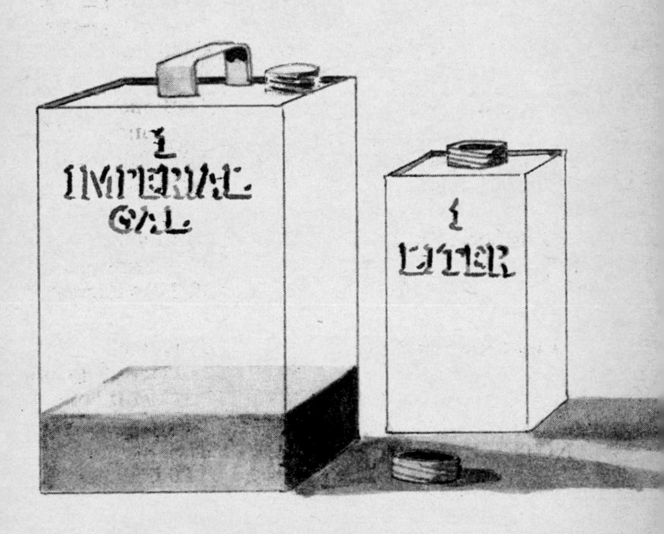

WEIGHT

WEIGHT
AVOIRDUPOIS*
U.S. TO METRIC

kilograms = kilos = kg.
U.S. pounds = lbs. = #

1 dram	=	27 11/32 grains
1 grain	=	0.002,285 ounce
1 ounce	=	16 drams
	=	437.5 grains
	=	28.349 grams
1 pound	=	16 ounces
	=	256 drams
	=	7,000 grains
	=	0.4536 kilogram
1 hundred weight	=	100 pounds
	=	45.36 kilograms
1 ton (short)	=	2,000 pounds
	=	907.2 kilograms

GROSS WEIGHT

1 hundred weight	=	112 pounds
	=	50.80 kilograms
1 ton (long)	=	2,240 pounds
	=	1,016.0 kilograms
1 ton (metric)	=	1,000 kilograms
	=	2,204.6 pounds

*NOTE: Unless otherwise specified all weights used in this publication are *Avoirdupois* weights.

METRIC TO U.S.

1 grain	=	0.002,285 ounce
1 milligram	=	0.015,43 grain
	=	0.000,035 ounce
1 carat	=	200 milligrams
	=	3.086 grains
	=	0.007,05 ounce
1 centigram	=	10 milligrams
	=	0.1543 grain
	=	0.000,35 ounce
1 decigram	=	10 centigrams
	=	1.543 grains
	=	0.003,52 ounce
1 gram	=	1,000 milligrams
	=	100 centigrams
	=	10 decigrams
	=	0.0352 ounce
1 dekagram	=	10 grams
	=	1,000 centigrams
	=	0.352 ounce
1 hectogram	=	100 grams
	=	10 dekagrams
	=	0.2204 pound
	=	3.52 ounces
1 kilogram	=	1,000 grams
	=	10 hectograms
	=	2.204 pounds
	=	35.274 ounces
1 ton metric	=	1,000 kilograms
	=	2,204.6 pounds

*NOTE: Unless otherwise specified all weights used in this publication are *Avoirdupois* weights.

GRAMS
METRIC TO U.S.

1,000,000 grams	=	1 millier
	=	1 cubic meter = $m.^3$
	=	2,204.6 pounds
100,000 grams	=	1 quintal
	=	1 hectoliter
	=	220.46 pounds
10,000 grams	=	1 myriagram
	=	10 liters
	=	22.046 pounds
1,000 grams	=	1 kilogram
	=	1 liter
	=	2.2046 pounds
100 grams	=	1 hectogram
	=	1 deciliter
	=	3.527 ounces
10 grams	=	1 dekagram
	=	10 cubic centimeters
	=	0.3527 ounce
1 gram	=	1 cubic centimeter
	=	15.432 grains
1/10 gram	=	1 decigram
	=	1/10 of a cubic centimeter
	=	1.543 grains
1/100 gram	=	1 centigram
	=	10 cubic millimeters
	=	0.154 grain
1/1,000 gram	=	1 milligram
	=	1 cubic millimeter
	=	0.0154 grain

WEIGHT

One U.S. Ounce = 28.35 grams

One Gram = .035,27 oz. (*avoirdupois*)

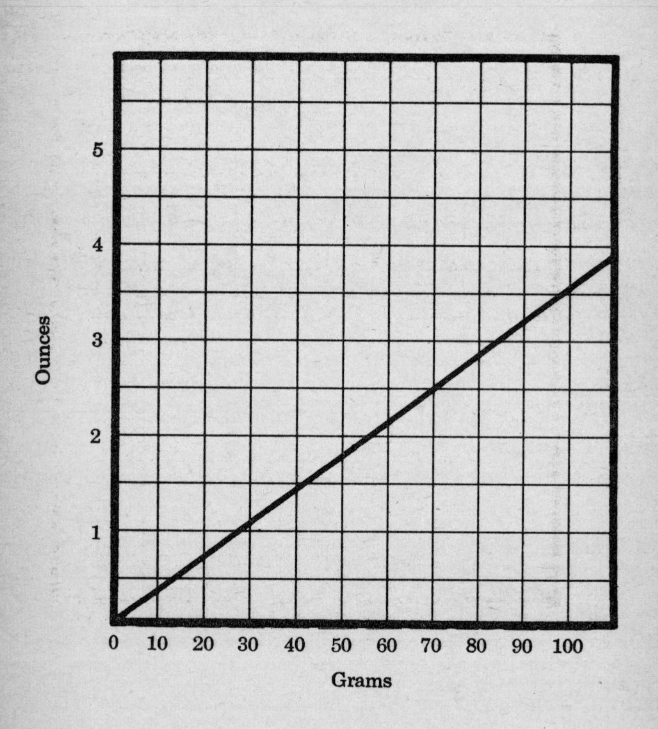

WEIGHT

One U.S. Ounce = 28.35 grams

One Gram = .035,27 oz. (*avoirdupois*)

One Kilogram (kg.) = 1,000 grams
= 35.27 ounces

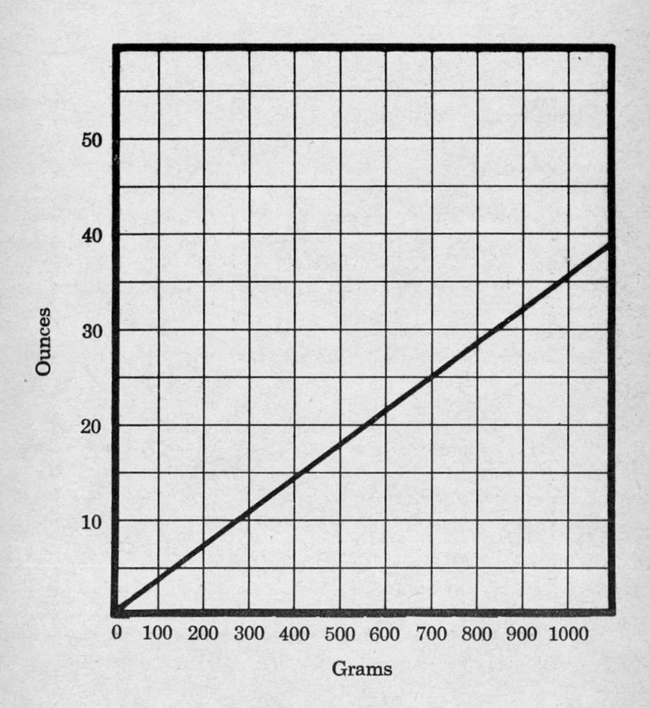

WEIGHT

One Kilogram = 2.2046 U.S. pounds

One U.S. Pound = 0.4536 kgs.

NOTE: kilograms = kilos = kgs.
pounds = lbs. = #

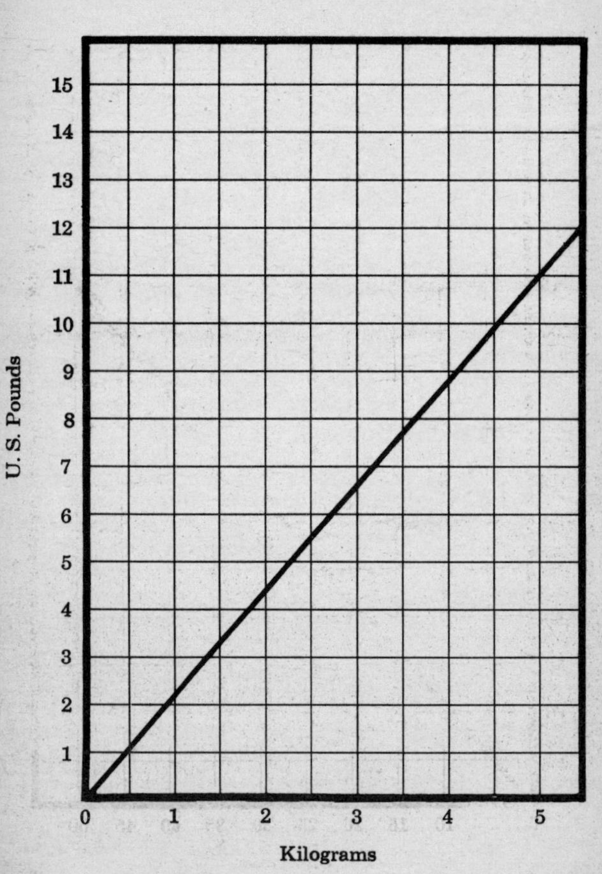

WEIGHT

One Kilogram = 2.2046 U.S. pounds

One U.S. Pound = 0.4536 kgs.

NOTE: kilograms = kilos = kgs.
pounds = lbs. = #

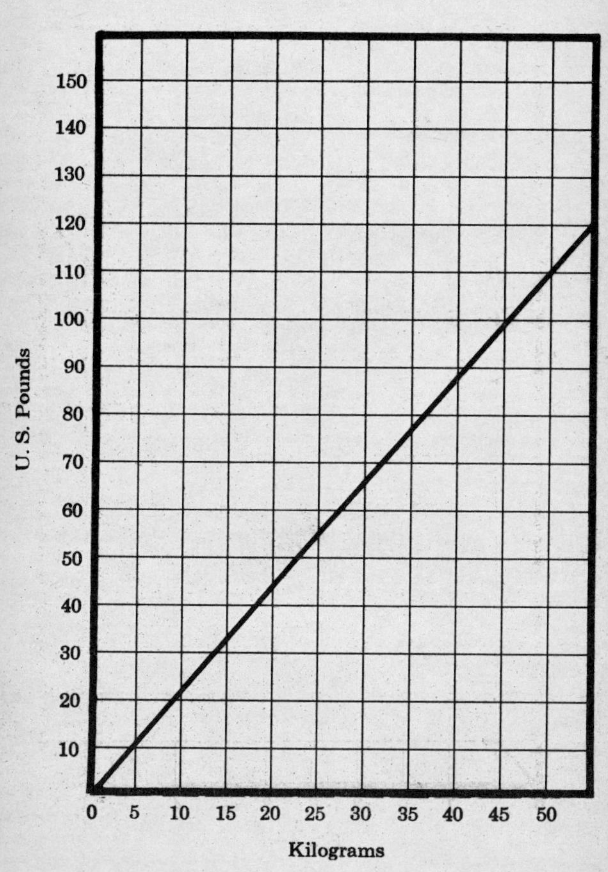

WEIGHT

One Kilogram = 2.2046 U.S. pounds

One U.S. Pound = 0.4536 kgs.

NOTE: kilograms = kilos = kgs.
pounds = lbs. = #

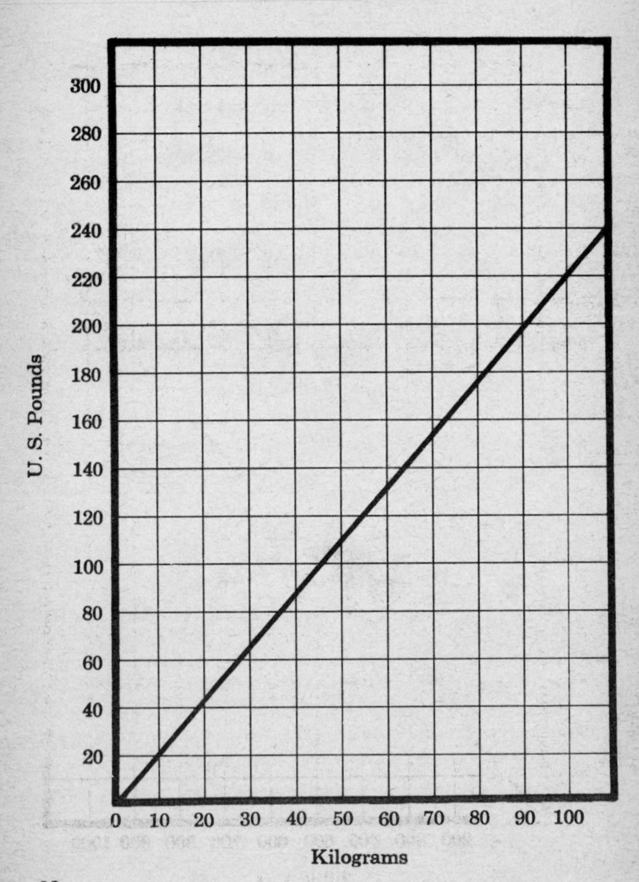

WEIGHT

One Kilogram = 2.2046 U.S. pounds

One U.S. Pound = 0.4536 kgs.

NOTE: kilograms = kilos = kgs.
pounds = lbs. = #

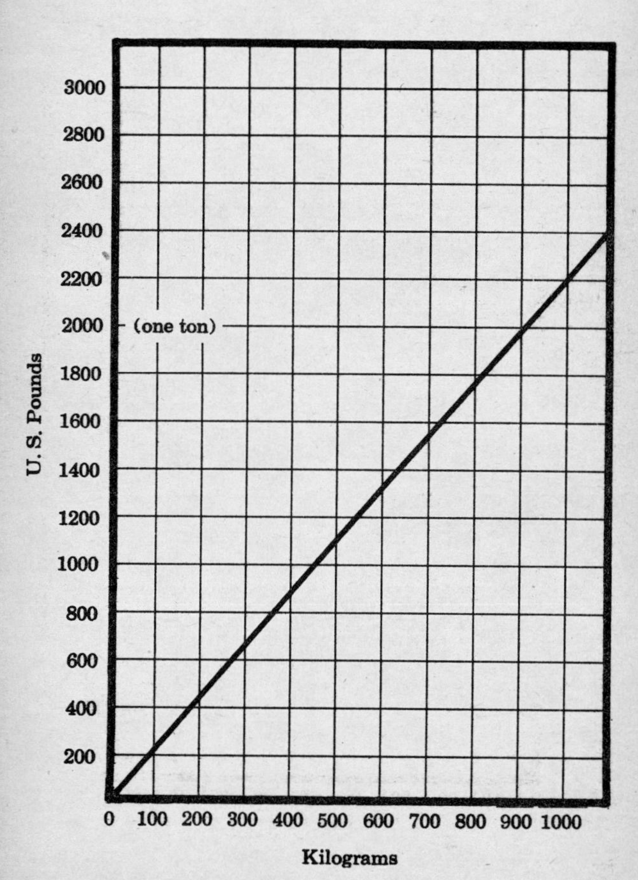

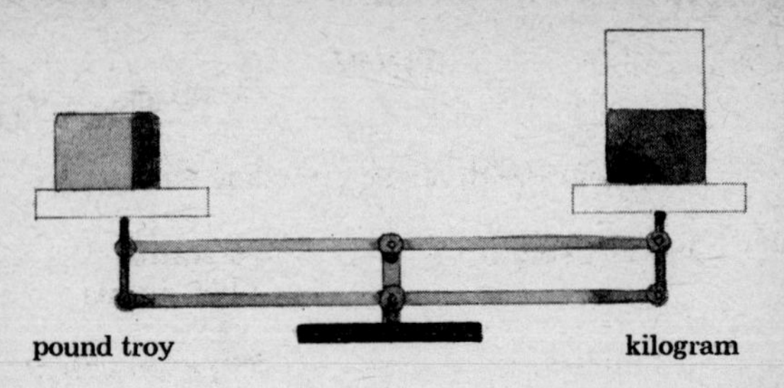
pound troy
kilogram

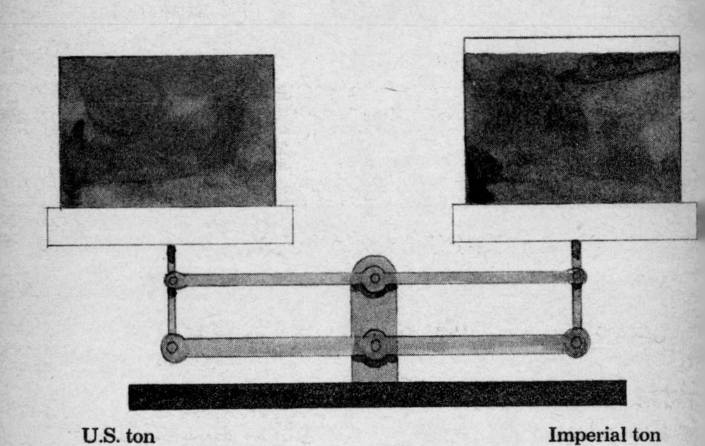
U.S. ton
Imperial ton

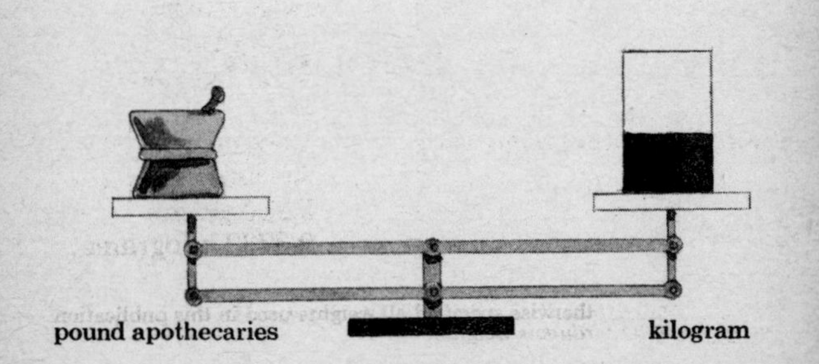
pound apothecaries
kilogram

TROY

(used for gold, silver, precious stones)

1 pennyweight	=	24 grains
	=	1.555 grams
1 ounce troy	=	20 pennyweights
	=	480 grains
	=	31.103 grams
1 pound troy	=	12 ounces troy
	=	240 pennyweights
	=	5,760 grains
	=	373.24 grams
	=	0.3732 kilogram

APOTHECARIES

(used in pharmacy)

1 scruple	=	20 grains
	=	1.296 grams
1 dram	=	3 scruples
	=	60 grains
	=	3.888 grams
1 ounce apothecaries	=	8 drams
	=	24 scruples
	=	480 grains
	=	31.103 grams
1 pound apothecaries	=	12 ounces
	=	96 drams
	=	5,760 grains
	=	0.3732 kilograms

NOTE: Unless otherwise specified all weights used in this publication are *Avoirdupois* weights.

BRITISH

1 ounce	=	28.349 grams
1 pound	=	16 ounces
	=	453.6 grams
	=	0.4536 kilogram
1 stone	=	14 pounds
	=	6.35 kilograms
1 quarter	=	2 stones
	=	28 pounds
	=	12.70 kilograms
1 hundred weight	=	4 quarters
	=	8 stones
	=	112 pounds
	=	50.80 kilograms
1 ton	=	20 hundred weight
	=	2,240 pounds
	=	1,016.0 kilograms

The British *Avoirdupois* pound, Troy pound, and Apothecaries pound are the same as the U.S. pound.

METRIC TERMINOLOGY

AREA

1 square kilometer ($km.^2$)	=	1,000,000 square meters
	=	100 hectares
1 hectare	=	10,000 square meters
	=	100 ares
1 are	=	100 square meters
1 square meter ($m.^2$)	=	1,000,000 square millimeters
	=	10,000 square centimeters
1 square centimeter ($cm.^2$)	=	100 square millimeters

NOTE:	1 $km.^2$	=	247.11 acres
		=	0.386 square mile
	1 $m.^2$	=	10.764 square feet
		=	1.196 square yards

LENGTH

1 kilometer (km.)	=	1,000 meters
	=	10 hectometers
1 hectometer	=	100 meters
	=	10 dekameters
1 dekameter	=	10 meters
1 meter	=	1,000 millimeters
	=	10 decimeters
1 decimeter	=	100 millimeters
	=	10 centimeters
1 centimeter	=	10 millimeters

NOTE:	1 km.	=	0.621 mile
		=	3,280.8 feet
	1 m.	=	39.37 inches

VOLUME, LIQUID CAPACITY

1 kiloliter	=	1,000 liters
	=	10 hectoliters
1 hectoliter	=	100 liters
	=	10 dekaliters
1 dekaliter	=	10 liters
1 liter	=	1,000 milliliters
	=	10 deciliters
1 deciliter	=	100 milliliters
	=	10 centiliters
1 centiliter	=	10 milliliters

NOTE:	1 liter	=	0.2642 U.S. gallons
	1 kiloliter	=	264.2 U.S. gallons

VOLUME, CUBIC MEASURE

1 cubic meter ($m.^3$)	=	1 stere
	=	1,000,000 cubic centimeters
	=	1,000 cubic decimeters
1 cubic decimeter	=	1,000,000 cubic millimeters
	=	1,000 cubic centimeters
1 cubic centimeter ($cm.^3$)	=	1,000 cubic millimeters

NOTE: 1 $m.^3$ = 35.31 cubic feet

WEIGHT

1 metric ton	=	1,000 kilograms
1 kilogram	=	1,000 grams
	=	10 hectograms
1 hectogram	=	100 grams
	=	10 dekagrams
1 dekagram	=	10 grams
1 gram	=	1,000 milligrams
	=	10 decigrams
1 decigram	=	100 milligrams
	=	10 centigrams
1 centigram	=	10 milligrams

NOTE:

1,000 kgs.	=	2204.6 pounds
1 kg.	=	2.2046 pounds
100 grams	=	3.527 ounces

PIPE CONNECTIONS

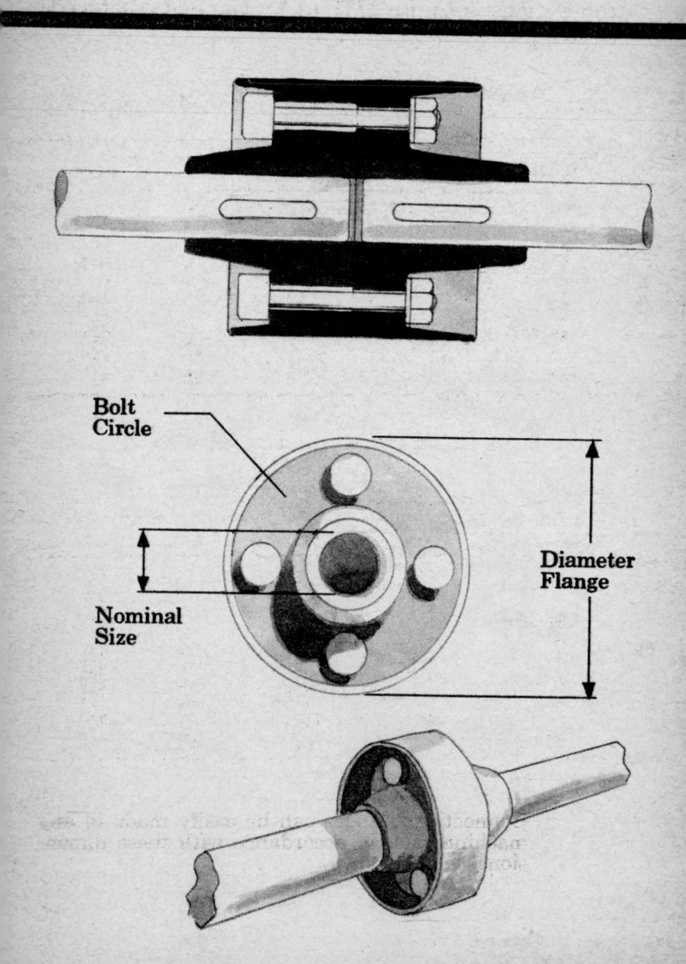

PIPE CONNECTIONS

The simplest most economical method of screwed connections is to use U.S. to Metric male or female adapters.

U.S. STANDARD FLANGE DIMENSIONS

Nominal Size	Diameter	Bolt Circle	Bolt No.	Bolt Size
1″	4¼″	3⅛″	4	½″
1½″	5″	3⅞″	4	½″
2″	6″	4¾″	4	⅝″
2½″	7″	5½″	4	⅝″
3″	7½″	6″	4	⅝″
4″	9″	7½″	8	⅝″
6″	11″	9½″	8	¾″
8″	13½″	11¾″	8	¾″
10″	16″	14¼″	12	⅞″
12″	19″	17″	12	⅞″

BRITISH STANDARD FLANGE DIMENSIONS

Nominal Size	Diameter	Bolt Circle	Bolt No.	Bolt Size
1″	4½″	3¼″	4	½″
1½″	5¼″	3⅞″	4	½″
2″	6″	4½″	4	⅝″
2½″	6½″	5″	4	⅝″
3″	7¼″	5¾″	4	⅝″
4″	8½″	7″	4	⅝″
6″	11″	9¼″	8	⅝″
8″	13¼″	11½″	8	⅝″
10″	16″	14″	8	¾″
12″	18″	16″	12	¾″

NOTE: Connecting Flanges can be easily made in any machine shop in accordance with these dimensions.

METRIC FLANGE DIMENSIONS (DIN)

Nominal Size		Outside Diameter		Bolt Circle		Bolts		Hole Diameter	
U.S. inches	DIN mm.	U.S. inches (Actual)	DIN mm. (Metric)	U.S. inches (Actual)	DIN mm. (Metric)	DIN No./Size mm.		U.S. inches	DIN mm.
1"	25	4.528	115	3.346	85	4	12	.551	14
1½"	40	5.905	150	4.331	110	4	16	.708	18
2"	50	6.496	165	4.921	125	4	16	.708	18
2½"	65	7.283	185	5.709	145	4	16	.708	18
3"	80	7.874	200	6.299	160	8	16	.708	18
4"	100	8.661	220	7.086	180	8	16	.708	18
6"	150	11.220	285	9.449	240	8	20	.905	23
8"	200	13.385	340	11.614	295	12	20	.905	23
10"	250	15.920	405	13.990	355	12	24	1.063	27
12"	300	18.100	460	16.150	410	12	24	1.063	27

CLOTHING

CLOTHING SIZES

MEN

HATS

U.S.	Metric
6½	52
6¾	54
7	56
7¼	58
7½	60

SHIRTS

U.S.	Metric
13	33
14	35
15	37
16	40
17	42

SHOES

U.S.	Metric
6	38
7	40
8	41
9	43
10	44
11	45
12	46

SOCKS

U.S.	Metric
9	23
10	25½
11	28
11½	29¼
12	30½

WOMEN

DRESSES

U.S.	English	French
10	32	38
12	34	40
14	36	42
16	38	44
18	40	46
20	42	48

SHOES

U.S.	English	Metric
4	2	34
5	3	35
6	4	36
7	5	38
8	6	38½
9	7	40
10	8	41

HATS

U.S.	Metric
21	53
22	56
23	58
24	61
24½	62

STOCKINGS

U.S.	Metric
8	20¼ (size 0)
9	22¾ (size 2)
10	25¼ (size 4)
11	27¾ (size 6)